NAVIGATING WATER QUALITY OUTCOMES IN AMERICAN WATERSHEDS

Whether due to climate change, drought, flooding, competing demands, or pollution, watersheds across the globe are under significant duress. To respond to these complex challenges, collaborative approaches to watershed governance have increasingly been adopted in the United States, but very few studies have yet to systematically assess their true effectiveness. This book addresses a significant gap in research by undertaking a comprehensive study of alternative, collaborative structures and whether these produce better water quality outcomes than traditional regulatory governance. Analyzing almost one quarter of US watersheds and examining both the revealed and perceived outcomes of watershed stakeholder collaboration, it is the first large-scale study on this topic. The insights the chapters provide will equip readers with a nuanced and generalizable understanding of the effectiveness of collaboration in natural resource management, which will be of great interest to researchers and practitioners in wide-ranging environmental and public policy roles.

DR. DAVID PRESCOTT ADAMS is Associate Professor of Public Administration and Master of Public Administration Director at California State University, Fullerton. His research examines collaborative governance, administrative discretion, and environmental policy, with an emphasis on watershed management, institutional resilience, and regulatory practice. He received Auburn University's Distinguished Dissertation Award in 2016 and has since published in *Policy Studies Journal, Environmental Policy & Governance, Administrative Theory & Praxis, Journal of Environmental Policy & Planning*, and *Journal of Politics and Policy*, among others. He is coeditor of *Ethics in Public Administration: Ethics, Corruption, and Public Policy* (2019, Cognella) and teaches courses in public administration theory, the policy process, and collaborative governance.

DR. JONATHAN MARC FISK is Associate Professor in the School of Public Affairs at the University of Utah, where he also serves as the Associate Director for the Master of Public Administration and Master of Public Policy programs. Prior to his appointment at the University of Utah, he served as Assistant Professor as well as Associate Professor of Political Science at Auburn University. He is also the Chairperson for the American Society for Public Administration's Section on Environmental and Natural Resources Administration. He has published widely on intergovernmental relations, as well as on environmental, water, and energy policy for more than a decade. He is the author or coauthor of five books, including *Oil and Gas Waste Governance: State Politics of Induced Seismicity* in Ohio and Oklahoma (2025, Routledge, with Zachary Mahafza, Joseph A. Aistrup, and Lorraine W. Wolf); *The Drought Dilemma: Pathways to Innovation for a Water-Stressed Future* (2024, Routledge, with Jonathan Farley and John C. Morris); *The Shale Renaissance: How Fracking Changed Pennsylvania in the Twenty-First Century* (2022, University of Pittsburgh Press, with Soren Jordan and A.J. Good); *Intergovernmental Relations: State and Local Challenges in the Twenty-First Century* (2022, Routledge); and *The Fracking Debate: Intergovernmental Politics of the Oil and Gas Renaissance, Second Edition* (2017, Routledge). In addition, he has published more than 40 peer-reviewed journal articles and book chapters in leading outlets such as *Policy Studies Journal, Administrative Theory & Praxis, Energy Policy, American Review of Public Administration, Environmental Policy & Governance, Public Works Management & Policy, Society & Natural Resources, Review of Policy Research*, and *State & Local Government Review*.

JOHN CHARLES MORRIS is Professor and Dean's Distinguished Research Fellow in the Department of Political Science at Auburn University, and Visiting Researcher at the University of Twente, Enschede, the Netherlands. He has studied environmental policy and water policy for more than 30 years and has published widely in public administration and public policy. Among other volumes, he is the coeditor of *Speaking Green with a Southern Accent: Environmental Management and Innovation in the South* (2010, Lexington Press), and *True Green: Executive Effectiveness in the U.S. Environmental Protection Agency* (2012, Lexington Press). His recent books include *The Case for Grassroots Collaboration: Social Capital and Ecosystem Restoration at the Local Level* (2013, Lexington Press, with William A. Gibson, William M. Leavitt, and Shana C. Jones); *Organizational Motivation for Collaboration: Theory and Evidence* (2019, Lexington, with Luisa M. Diaz-Kope); *Interorganizational Interactions for Watershed Protection: Working Better Together* (2020, Routledge, with Madeleine W. McNamara); and *Clean Water Policy and State Choice: Promise and Performance in the Water Quality Act* (2022, Cambridge University Press). His most recent book is *The Drought Dilemma: Pathways to Innovation for a Water-Stressed Future* (2024, Routledge, with Jonathan D. Farley and Jonathan M. Fisk). In addition, he has published more than 95 articles in peer-reviewed journals and more than 40 book chapters, reports, and other publications.

NAVIGATING WATER QUALITY OUTCOMES IN AMERICAN WATERSHEDS

DAVID PRESCOTT ADAMS
California State University, Fullerton

JONATHAN MARC FISK
University of Utah

JOHN CHARLES MORRIS
Auburn University

CAMBRIDGE
UNIVERSITY PRESS

CAMBRIDGE
UNIVERSITY PRESS

Shaftesbury Road, Cambridge CB2 8EA, United Kingdom

One Liberty Plaza, 20th Floor, New York, NY 10006, USA

477 Williamstown Road, Port Melbourne, VIC 3207, Australia

314–321, 3rd Floor, Plot 3, Splendor Forum, Jasola District Centre,
New Delhi – 110025, India

Cambridge University Press is part of Cambridge University Press & Assessment,
a department of the University of Cambridge.

We share the University's mission to contribute to society through the pursuit of education, learning and research at the highest international levels of excellence.

www.cambridge.org
Information on this title: www.cambridge.org/9781009567749
DOI: 10.1017/9781009567725

When citing this work, please include a reference to the DOI 10.1017/9781009567725

First published 2026

A catalogue record for this publication is available from the British Library

Library of Congress Cataloging-in-Publication Data
Names: Adams, David Prescott author | Fisk, Jonathan Marc author |
Morris, John Charles author
Title: Navigating water quality outcomes in American watersheds / David
Prescott Adams, Jonathan Marc Fisk, John Charles Morris.
Description: First edition. | New York : Cambridge University Press, 2026.
| Includes bibliographical references and index.
Identifiers: LCCN 2025048406 (print) | LCCN 2025048407 (ebook) |
ISBN 9781009567749 hardback | ISBN 9781009567725 ebook
Subjects: LCSH: Water quality – United States | Watersheds – United States
Classification: LCC TD223 .A466155 2026 (print) | LCC TD223 (ebook)
LC record available at https://lccn.loc.gov/2025048406
LC ebook record available at https://lccn.loc.gov/2025048407

ISBN 978-1-009-56774-9 Hardback

To Anne M. Strand
– David Prescott Adams

To Melanie, Noah, Micah, and Caleb
– Jonathan Marc Fisk

To Anne James
– John Charles Morris

Contents

Figures

Tables

Acknowledgments

We are deeply grateful to our families and loved ones for their patience, encouragement, and unwavering support throughout the development of this book. We also extend our thanks to our colleagues and home institutions – the Division of Politics, Administration, and Justice at California State University, Fullerton; the Department of Political Science in the College of Liberal Arts at Auburn University; and the School of Public Affairs at the University of Utah – for fostering an intellectual community and providing the space to pursue this work.

We are especially thankful to the editorial team at Cambridge University Press for their guidance, professionalism, and commitment to bringing this project to publication. Finally, we extend our appreciation to the anonymous reviewers, whose careful feedback and critical insights strengthened the manuscript considerably.

Abbreviations

AIC	Akaike Information Criterion
ATTAINS	Assessment, TMDL Tracking and Implementation System
BIC	Bayesian Information Criterion
BIMBY	Because It's My Back Yard
BMP	Best Management Practices
BRM	Binary Response Model
CEC	Contaminants of Emerging Concern
CIP	Capital Improvement Plan
CPR	Common Pool Resource
CSO	Combined Sewer Overflow
CWA	Clean Water Act
CWSRF	Clean Water State Revolving Fund
DV	Dependent Variable
ENR	Environmental and Natural Resources
EPA	Environmental Protection Agency
GAO	Government Accountability Office
GIS	Geographic Information System
HUC	Hydrologic Unit Code
IAD	Institutional Analysis and Development
ICC	Intraclass Correlation Coefficient
LCAF	Levels of Collaboration Action Framework
MOU	Memorandum of Understanding
NDVI	Normalized Difference Vegetation Index
NEP	National Estuary Program
NGO	Nongovernmental Organization
NOAA	National Oceanographic and Atmospheric Administration
NPDES	National Pollution Discharge Elimination System
NPS	Nonpoint Source

OLS	Ordinary Least Squares
OSDS	On-Site Domestic Sewage
PCB	Polychlorinated Biphenyls
PPCP	Pharmaceutical and Personal Care Products
PPP	Public–Private Partnership
SES	Socioeconomic Status
US	United States
USDA	US Department of Agriculture
USGS	US Geological Survey
WQA	Water Quality Act
WQS	Water Quality Standards

Introduction

Over the past three decades, the rise of grassroots environmental collaborations represents a significant shift in American environmental governance, particularly in natural resource management (Lubell et al. 2002). This change is especially evident in watershed partnerships, where governance models have increasingly moved away from traditional, command-and-control and centralized approaches toward decentralized, comanaged, and collaborative structures. These partnerships typically involve diverse, multisector stakeholders who collaborate to direct resources and activities across both vertical and horizontal dimensions in an open, participative framework (Morris, et al. 2013; Genskow and Born 2006; Gerlak and Heikkila 2007; Imperial 2005). Today, watershed partnerships operate along a continuum of collaborative arrangements, ranging from highly formal processes coordinated among government agencies to informal, locally managed initiatives led by citizen groups (Mayer and Kenter 2016; McNamara 2016; Hardy and Koontz 2009; Koontz and Johnson 2004; Leach and Pelkey 2001; Margerum 2008). The nature of watersheds themselves is conducive to such approaches; watersheds often span political, economic, and social boundaries (Kenney et al. 2000; McGinnis 1999; McKinney and Harmon 2004), presenting challenges for the management of common-pool resources.

The prodigious growth of the use of watershed partnerships in the United States is reshaping the landscape of democratic environmental governance. These alternative governance structures are a significant shift in how governments and citizens interact, with both working together to address complex environmental issues (Kettl 2002; Sirianni 2009). Watershed partnerships have also garnered attention from scholars in fields ranging from ecology and environmental law to sociology, political science, and public administration (Goldfarb 1994; Weber 2000; Kenney 2000; Leach and Pelkey 2001; Sirianni 2009). More importantly, the growth of watershed partnerships raises several important research questions: How are these partnerships structured? What are the important features of their processes? How are they different from more traditional "command-and-control" (regulatory) approaches to governance? And, perhaps most importantly, do they produce results that are equal to, or better than, more traditional governance approaches? This last question is the focus of this book.

There is significant evidence in the extant literature that watershed partnerships produce positive results. Sabatier et al. (2005) and Morris et al. (2013), for instance, report robust results from citizen-based watershed collaboration, findings echoed elsewhere in the literature. There is also debate about the underlying reasons why collaboration is successful (see Weber et al. 2024; Diaz-Kope and Morris 2019). On the basis of these reports, one might be tempted to conclude that watershed collaboration is always successful. However, while we seem to know much about the "success" factors for watershed collaboration, less is actually known about why collaboration fails (McNamara et al. 2020).

There is another, equally important, limitation in the literature on watershed collaboration. In short, nearly all of the existing studies of watershed collaboration that examine results of collaboration are small-*n* case studies. Most studies tend to focus on fewer than five cases (see Sabatier et al. 2005; Morris et al. 2013); other studies (e.g., McNamara and Morris 2021; McNamara et al. 2010; Diaz-Kope and Morris 2022; 2019) address only one case. Even research that aggregates findings from other studies (see, e.g., Duram and Brown 1999) is limited by a relatively small number of cases. The findings in these studies, however, are remarkably consistent: collaboration works. Whereas there is variation in the explanations of *why* collaboration works (see McNamara et al. 2020), there is rather less variation in why these scholars chose the cases they did – convenience. Furthermore, nearly all of these studies are in-depth case studies. The case study is exceptionally good at description and explanation (see Yin 1994) but suffers from a lack of generalizability.

A case in point is the Morris et al. (2013) study of watershed collaboration in the Hampton Roads, Virginia area. Morris and colleagues examined the three watershed collaboration groups active in the region, each in a different watershed. The three cases were similar in that they were all initiated by private citizens, but differed in regard to the goals of the organizations, the nature of the issues in the watershed, and the composition and structure of the membership. Their study found plenty of similarities between the cases, but the differences make generalizability even within the region – not to mention the state or the larger Chesapeake Bay watershed – practically impossible.

This general approach in the watershed collaboration literature suggests the possibility of unintentional (or intentional) bias; scholars may be choosing cases that reflect their preconceived notions about the efficacy of collaboration. While we make no assertions about the motivations for any study, there is clearly a gap in our understanding of the effectiveness of watershed collaboration in terms of measurable changes in a watershed.[1] In short, we need better, and more generalizable, evidence that collaboration is an effective governance structure for addressing watershed protection. Lacking such evidence, we are resigned, at best, to a limited theoretical understanding of watershed collaboration.

Outputs vs. Outcomes in Collaboration Efforts

An important element of any discussion of the effects of collaboration revolves around the distinction between outputs and outcomes. These terms are often employed interchangeably or, perhaps even more often, without definition. If lumped into a single category, these terms can be generically thought of as "the things that collaboration produces." Analytically, though, to treat these terms or efforts generically substantially limits our understanding of the effects of collaboration and undervalues the potential benefits of collaboration. To that end, we turn our attention to the development of a more nuanced understanding of these terms.

In the watershed literature, Koontz and Thomas (2006, 115) were among the first to draw a distinction between these terms by defining collaboration outputs as the "intermediary causal mechanisms between collaboration process and collaborative outcomes." Indeed, that same body of literature treats outputs as the main variable of interest due to the costs of long-term monitoring and the problem of confounding variables (Mandarano 2008; Koontz and Thomas 2006). Morris et al. (2013) make a similar point, arguing that outputs are more immediate and thus more attractive as measures of success even if outcomes are better suited to measure progress toward goals.

Common examples of watershed collaborative outputs include project plans and agreements, new (or updated) water quality standards, scientific reports (Lubell et al. 2005; Mandarano 2008), or even increased social capital (Morris et al. 2013). Plans and agreements differ in that plans include input from stakeholders, choices around issue framing, and the nature of the setting (Duram and Brown 1999), whereas agreements represent the implementation phase (Leach, Pelkey, and Sabatier 2002). Agreements can range from broad outlines of objectives to specific action plans. Studies suggest that the level of sophistication of agreements may depend on the particular mix of stakeholders and their individual agendas (Hardy and Koontz 2009; Margerum 2008; Imperial 2005). Moreover, the ability

[1] A related issue is whether we can, or should, evaluate the effectiveness of watershed collaboration on the basis of measurable changes in a watershed. There are multiple problems with this assumption, not the least of which are the long time horizon necessary to detect changes, and the problem of quality data regarding water quality changes in a watershed. These issues will be addressed later in this book.

to arrive at actionable agreements may serve to increase social capital in collaborative settings (Morris et al. 2013; Trachtenberg and Focht 2005). As a result, many studies employ surrogate measures to assess outputs, such as the number of outreach activities, the number of restoration activities completed, the number of education programs delivered (Morris et al. 2013; Campbell, Koontz, and Bonnell 2011; Margerum 2007; Lubell et al. 2002).

It is also important to note that outputs, as defined previously, are unlikely to produce measurable changes in watershed quality. Rather, outputs are more effective as an indicator of the success of the collaborative process itself, as opposed to a measure of the environmental goals of the collaboration (Imperial 2005). Lubell (2004), among others, argues that linking changes in watershed quality to the activities of grassroots collaborative efforts is, at best, tenuous. Leach, Pelkey, and Sabatier (2002) note that many grassroots collaborations lack the resources, including the scientific expertise, to properly assess physical changes in the watershed. However, Morris et al. (2013) observed at least one case in which changes were not only apparent but also consistent with the overall goals of the group. In the case of Lynnhaven River NOW the goal was to be able to eat oysters from the Lynnhaven River. The water quality measurements were undertaken by the state's health department, and when the waters were deemed safe enough for human consumption of oysters, farmed oysters were harvested and consumed. The group still celebrates this accomplishment every year by hosting an annual oyster roast with oysters harvested from the river (Morris et al. 2013).

However, watershed outcomes can be conceived of as long-term changes in a watershed. If outputs are projects, then outcomes are the long-term effects of those projects; if outputs are education programs, the outcomes are changes in behaviors resulting from the education programs. Recent scholarship generally divides outcomes into two categories: environmental and social outcomes (see Koontz and Thomas 2006). Environmental outcomes are measurable changes in ecological systems such as improved habitats or improved water quality; social outcomes involve measurable changes in social structures such as volunteerism or civic engagement (Morton 2008; Thomas 2008). While some scholars treat trust (or social capital) as an outcome, Morris et al. (2013) treat social capital as a specific category of output and suggest that social capital, as a product of collaborative processes, offers a much more immediate effect.

The literature also distinguishes two important elements related to the effectiveness of environmental outcomes. The first of these is the implementation of projects designed to accomplish things such as nonpoint pollution abatement, changes to in-stream flows, the establishment (or restoration) of riparian zones. The second category involves changes in land use through zoning, easements, or acquisition (Leach, Pelkey, and Sabatier 2002). Mandarano (2008) also argues that changes in ecosystem biodiversity are reasonable measures of effectiveness, while Conley and Moote (2005) suggest that changes in the amount or frequency of environmental litigation are an effective outcome measure. Still others suggest indicators such as changes in behaviors, intellectual capital, environmental ethics, and perceptions of conditions can be employed as measures of outcomes (McNamara, Leavitt, and Morris 2010; McKinney and Harmon 2004; Weber 2009).

Previous studies also suggest that the perceptions of stakeholder participants about changes in watershed quality may be biased (see Koontz and Thomas 2006). This may be due, in part, to the solidarity (or social capital) that develops as a result of the process itself, with participants exhibiting a form of confirmation bias (it's successful because we're a part of it). The net effect is a "halo bias" that becomes more important to participants than substantive scientific evidence (Koontz and Thomas 2006; Trachtenberg and Focht 2005). We also note that such evaluations require substantially less evidence, are less expensive than scientific evidence, and do not require long time horizons (see Sabatier, Weible, and Ficker 2005; Kenney et al. 2000). Finally, we also note that there is a dearth of studies that have attempted to compare participant perceptions to more scientific measurements of changes in watershed quality. One of the goals of this study is to provide such a comparison to determine the degree of mismatch between perceptions and empirical evidence.

For the purposes of this study, we distinguish between outputs and outcomes in terms of time, where the temporal "space" between the concepts differs. Because implementation begins at the point at which resources are assigned and employed to a project or plan, and typically occurs in the near term, we believe these activities are better thought of as outputs. Generally speaking, outcomes are defined as long-term changes in environmental conditions, and may not be obvious (or measurable) in the near term. While we recognize the challenges in measuring outcomes, the water quality data we employ in this study are better suited to a role as outcome measures. Subject to the limitations of these data, we believe the water quality data is the best available across a range of geographic locations (and over time), and are thus useful as measures of outcome measures of water quality.

Finally, we must note that there are cases in which water quality may improve in a short period of time. Morris et al. (2013) report the results of a project in the Elizabeth River in Norfolk, Virginia, that removed thousands of cubic yards of contaminated riverbed. Within weeks of the completion of the project, scientists from the Virginia Institute of Marine Science noted a remarkable improvement in both the health and numbers of marine species in that segment of the river. Water samples taken in the area also saw significant reductions in the pollutants removed from the area. While this case is noteworthy in its own right, we suggest these kinds of results are more likely indicative of outlier cases.

Goals of the Research

The point of this book is, then, to offer a study of collaboration outputs and outcomes that examines a large number of cases across as many locations as possible. By examining many cases, we are better able to draw generalizable conclusions about the efficacy of collaboration as a mechanism to achieve water quality goals. Moreover, by examining both the perceptions of collaboration participants and empirical water quality testing data to examine changes in the watershed, we are able to speak directly to the effectiveness of collaboration as a governance tool to improve watershed quality.

We must also note that the standard for "success" cannot rely simply on either empirical or perceptual data. Indeed, it is possible that these two data points are measuring slightly different things. As Morris et al. (2013) pointed out, an important outcome of collaborative processes is an increased level of social capital; they also noted that empirical improvements in a watershed may take years to manifest. Likewise, as we will discuss later in this volume, there are significant issues with the empirical water quality data that states submit to the Environmental Protection Agency (EPA) and that the EPA then makes available on its website. However, these data are the best available and still offer a comparative measure to determine water quality improvements.

With this in mind, we also note that the development of a comparison group (or a control group), while optimal, is nearly impossible in this setting. First, there is no mechanism to identify cases in which efforts are made to address watershed quality that do not involve collaboration. Such efforts exist, but to identify specific projects would require a herculean effort, as would identifying the entities primarily responsible in each of those cases. Additionally, collaboration is widespread enough that such projects likely overlap (and are likely coexisting) with other efforts, which would make it difficult or impossible to separate the effects of one from the other. So, while we can say that water quality improves (or not) in settings where collaboration is employed, we cannot make claims about whether collaboration is more (or less) effective as a governance structure than other structures for addressing water quality. Thus, the contribution of this work is to offer a large-*n* study of watershed collaboration, and to determine whether, on the whole, watershed quality improves where collaborative efforts are employed. This moves us beyond the more fragmented, atomistic, non-generalizable (still valuable!) case studies already present in the literature, and serves to provide a benchmark against which the claims of success in those cases can be measured.

Overview of This Book

This book explores the use of collaboration as a mechanism to address real and perceived results associated with watershed management, protection, and restoration. As noted at the outset of this introduction, we are focused on the ability of these collaborative efforts to have a measurable impact on the watersheds on which they expend their efforts. Unlike previous studies, we do not rely on case studies; rather, we offer a more comprehensive examination of watershed collaboration efforts from across the United States. We supplement self-reported evaluations from the groups with publicly collected and publicly available data on water quality and watershed improvements to determine whether the ecosystem has improved.

This book is divided into eight chapters. Chapter 1 introduces the specific research questions and makes a case for why this study matters – both in terms of collaboration theory and practice. Chapter 2 summarizes the emergence and evolution of water governance in the US. This includes both the formal adoption of key environmental laws such as the Clean Water Act of 1972 through compliance and related challenges. Here, collaboration, as an alternative to traditional regulation, is situated in its broader political, socio-technical, and legal context. In Chapter 3, we summarize the relevant literature with an emphasis on key dimensions and characteristics associated with effective forms of collaboration. The chapter also presents a model of watershed collaboration that seeks to explain both outputs and outcomes (both revealed and perceived). Relying on data from 306 watersheds, Chapters 4 and 5 evaluate the model with an eye toward *revealed* water quality outputs and outcomes. Chapters 6 and 7 look toward *perceptions* of water quality via a survey of 114 unique watersheds. Chapter 7 also leverages the structure of the Institutional Analysis and Development framework to weave together the voices of watershed actors stakeholders. Respondents here reflect on their central values and goals, operational activities and outputs, and the role of interpersonal relationships and trust. Chapter 8 concludes with lessons for theory, practice, and future research in collaborative environmental governance.

References

Agranoff, Robert. 2006. Inside Collaborative Networks: Ten Lessons for Public Managers. *Public Administration Review*, **66**(Special Issue), 56–63.

Agranoff, Robert. 2007. *Managing within Networks: Adding Value to Public Organizations*. Washington, D.C.: Georgetown University Press.

Agranoff, Robert. 2008. Collaboration for Knowledge: Learning from Public Management Networks. In *Big Ideas in Collaborative Public Management*, edited by Lisa Blomgren Bingham and Rosemary O'Leary, 162–194. Armonk, NY: M.E. Sharpe.

Agranoff, Robert, and McGuire, Michael. 2003. *Collaborative Public Management*. Washington, D.C.: Georgetown University Press.

Bryson, John, Crosby, Barbara, and Stone, Melissa M. 2006. The Design and Implementation of Cross-Sector Collaborations: Propositions from the Literature. *Public Administration Review*, **66**, 44–53.

Campbell, Joseph T., Koontz, Tomas M., and Bonnell, Joseph E. 2011. Does Collaboration Promote Grass-Roots Behavior Change? Farmer Adoption of Best Management Practices in Two Watersheds. *Society and Natural Resources*, **24**, 1127–1141.

Conley, Alexander, and Moote, Margaret. 2003. Evaluating Collaborative Natural Resource Management. *Society and Natural Resources*, **16**, 371–386.

Cooper, Terry, Bryer, Thomas, and Meek, Jack. 2006. Citizen-Centered Collaborative Public Management. *Public Administration Review*, **66**(Special Issue), 76–88.

Diaz-Kope, Luisa, and Morris, John C. 2019. *Organizational Motivation for Collaboration: Theory and Evidence*. Lanham, MD: Lexington Press.

Diaz-Kope, Luisa, and Morris, John C. 2022. Why Collaborate? Exploring the Role of Organizational Motivations in Cross-Sector Watershed Collaboration. *Politics & Policy*, **50**(3), 516–539.

Diaz-Kope, Luisa, Miller-Stevens, Katrina, and Morris, John C. 2014. Collaboration Processes and Institutional Structures: Reexamining the Black Box. *International Journal of Public Administration*, **38**(9), 607–615.

Duram, Leslie A., and Brown, Katharin G. 1999. Assessing Public Participation in US Watershed Planning. *Society and Natural Resources*, **12**, 455–467.

Feiock, Richard C., and Scholz, John T. (eds.). 2010. *Self-Organizing Federalism: Collaborative Mechanisms to Mitigate Institutional Collective Action Dilemmas*. New York: Cambridge University Press.

Genskow, Kenneth D., and Born, Stephen M. 2006. Organizational Dynamics of Watershed Partnerships: A Key to Integrated Water Resources Management. *Journal of Contemporary Water Research & Education*, **35**, 56–64.

Gerlak, Andrea, and Heikkila, Tanya. 2007. Collaboration and Institutional Endurance in U.S. Water Policy. *Political Science and Politics*, **40**, 55–60.

Goldfarb, William. 1994. Watershed Management: Slogan or Solution? *Boston College Environmental Affairs Law Review*, **21**, 483–510.

Gray, Barbara. 1985. Conditions Facilitating Interorganizational Collaboration. *Human Relations*, **38**, 911–936.

Gray, Barbara. 1989. *Collaborating: Finding Common Ground for Multiparty Problems*. San Francisco: Jossey-Bass Inc.

Hale, Kathleen. 2011. *How Information Matters: Networks and Public Policy Innovation*. Washington, D.C.: Georgetown University Press.

Hardy, Scott D., and Koontz, Tomas. 2009. Rules for Collaboration: Institutional Analysis of Group Membership and Levels of Action in Watershed Partnership. *Policy Studies Journal*, **37**, 393–414.

Imperial, Mark. 2005. Using Collaboration as a Governance Strategy: Lessons from Six Watershed Management Programs. *Administration & Society*, **37**, 281–320.

Keast, Robyn, Mandell, Myrna P., Brown, Kerry, and Woolcock, Geoffrey. 2004. Network Structures: Working Differently and Changing Expectations. *Public Administration Review*, **64**(3), 363–370.

Kelley, Amy P., and Morris, John C. 2010. Keeping the Vision Alive: The Role of Networks in National Memorial Building. *Public Voices*, **11**(1), 89–109.

Kenney, Douglas. 2000. *Arguing about Consensus: Examining the Case against Western Watershed Initiatives and Other Collaborative Groups Active in Natural Resources Management*. Boulder, CO: Natural Resources Law Center, University of Colorado School of Law.

Kenney, Douglas S., McAllister, Sean T., Caile, William H., and Peckham, Jason S. 2000. *The New Watershed Source Book*. Boulder: Natural Resources Law Center.

Kettl, Donald F. 2002. *The Transformation of Governance: Public Administration for Twenty-First Century America*. Baltimore: The Johns Hopkins University Press.

Koontz, Tomas, and Johnson, Elizabeth. 2004. One Size Does Not Fit All: Matching Breadth of Stakeholder Participation to Watershed Group Accomplishments. *Policy Sciences*, **37**, 185–204.

Koontz, Tomas, and Thomas, Craig. 2006. What Do We Know and Need to Know about the Environmental Outcomes of Collaborative Management? *Public Administration Review*, **66**(Special Issue), 111–121.

Leach, William, and Pelkey, N. 2001. Making Watershed Partnerships Work: A Review of the Empirical Literature. *Journal of Water Resources Planning and Management*, **127**, 378–385.

Leach, William, Pelkey, N., and Sabatier, Paul. 2002. Stakeholder Partnerships as Collaborative Policymaking: Evaluation Criteria Applied to Watershed Management in California and Washington. *Journal of Policy Analysis and Management*, **21**, 645–670.

Lubell, Mark. 2004. Collaborative Watershed Management: A View from the Grassroots. *Policy Studies Journal*, **32**, 341–361.

Lubell, Mark, Schneider, Mark, Scholz, John T., and Mete, Mihriye. 2002. Watershed Partnerships and the Emergence of Collective Action Institutions. *American Journal of Political Science*, **46**, 148–163.

Lubell, Mark, Sabatier, Paul A., Vedlitz, Arnold, Focht, Will, Trachtenberg, Zev, and Matlock, Marty. 2005. Conclusions and Recommendations. In *Swimming Upstream: Collaborative Approaches to Watershed Management*. edited by Paul A. Sabatier, Will Focht, Mark Lubell, Zev Trachtenberg, Arnold Vedlitz and Marty Matlock, 261–296. Cambridge, MA: MIT Press.

Mandarano, Lynn. 2008. Evaluating Collaborative Environmental Planning Outputs and Outcomes: Restoring and Protecting Habitat and the New York-New Jersey Harbor Estuary Program. *Journal of Planning Education and Research*, **27**, 456–468.

Margerum, Richard D. 2007. Overcoming Locally Based Collaboration Constraints. *Society and Natural Resources*, **20**, 135–152.

Margerum, Richard D. 2008. A Typology of Collaboration Efforts in Environmental Management. *Environmental Management*, **41**, 487–500.

Martz, Robert J., McCarthy, William, and Morris, John C. 2018. The Consent Order as Mandated Collaboration: The Case of the Hampton Roads Sanitation District. *Public Works Management & Policy*, **23**(2), 168–185.

Mayer, Martin K., and Kenter, Robert C. 2016. The Prevailing Elements of Public Sector Collaboration. In *Advancing Collaboration Theory: Models, Typologies, and Evidence*, edited by John C. Morris and Katrina Miller-Stevens, 43–64. New York: Routledge.

McGinnis, Michael V. 1999. Making the Watershed Connection. *Policy Studies Journal*, **27**, 497–501.

McKinney, Matthew, and Harmon, W. 2004. *The Western Confluence: A Guide to Governing Natural Resources*. Washington, D.C.: Island Press.

McNamara, Madeleine W. 2012. Struggling to Untangle the Web of Cooperation, Coordination, and Collaboration: A Framework for Public Managers. *International Journal of Public Administration*, **35**(6), 389–401.

McNamara, Madeleine W. 2016. Unraveling the Characteristics of Mandated Collaboration. In *Advancing Collaboration Theory: Models, Typologies, and Evidence*, edited by John C. Morris and Katrina Miller-Stevens, 65–86. New York: Routledge.

McNamara, Madeleine W., and Morris, John C. 2012. More than a One-Trick Pony: Exploring the Contours of a Multi-Sector Convener. *Journal for Nonprofit Management*, **15**(1), 84–103.

McNamara, Madeleine W., and Morris, John C. 2021. *Multiorganizational Arrangements for Watershed Protection: Working Better Together*. New York: Routledge.

McNamara, Madeleine W., Leavitt, William M., and Morris, John C. 2010. Multiple Sector Partnerships and the Engagement of Citizens in Social Marketing Campaigns. *Virginia Social Science Journal*, **45**, 1–20.

McNamara, Madeleine W., Miller-Stevens, Katrina, and Morris, John C. 2020. Exploring the Determinants of Collaboration Failure. *International Journal of Public Administration*, **43**(1), 49–59.

Morris, John C., and Miller-Stevens, Katrina. 2016. The State of Knowledge in Collaboration. In *Advancing Collaboration Theory: Models, Typologies, and Evidence*, edited by John C. Morris and Katrina Miller-Stevens, 3–13 New York: Routledge.

Morris, John C., Gibson, William A., Leavitt, William M., and Jones, Shana C. 2013. *The Case for Grassroots Collaboration: Social Capital and Ecosystem Restoration at the Local Level*. Lanham, MD: Lexington Books.

Morton, Lois. 2008. The Role of Civic Structure in Achieving Performance-Based Watershed Management. *Society and Natural Resources*, **21**, 751–766.

Sabatier, Paul, Weible, Christopher, and Ficker, Jared. 2005a. Eras of Water Management in the United States: Implication for Collaborative Watershed Approaches. In *Swimming Upstream: Collaborative Approaches to Watershed Management*, edited by Paul Sabatier, W. Focht, M. Lubell, Z. Trachtenberg, A. Vedlitz and M. Matlock, 23–51. Cambridge, MA: MIT Press.

Sabatier, Paul A., Focht, Will, Lubell, Mark, Trachtenberg, Zev, Vedlitz, Arnold, and Matlock, Marty (eds.). 2005b. *Swimming Upstream: Collaborative Approaches to Watershed Management*. Cambridge, MA: The MIT Press.

Scholz, John T., Berardo, Ramiro, and Kile, Brad. 2008. Do Networks Solve Collective Action Problems? Credibility, Search, and Collaboration. *The Journal of Politics*, **70**(2), 393–406.

Sirianni, Carmen. 2009. The Civic Mission of a Federal Agency in the Age of Networked Governance: U.S. Environmental Protection Agency. *American Behavioral Scientist*, **52**, 933–952.

Thomas, Craig. 2008. Evaluating the Performance of Collaborative Environmental Governance. In: *Consortium on Collaborative Governance Mini-Conference*. University of Washington, Santa Monica.

Trachtenberg, Zev, and Focht, Will. 2005. Legitimacy and Watershed Collaborations. Pages 53–82 of: Sabatier, Paul, Focht, Will, Lubell, Mark, Trachtenberg, Zev, and Matlock, Marty (eds.), *Swimming Upstream: Collaborative Approaches to Watershed Management*. Cambridge, MA: MIT Press.

Weber, Edward. 2000. A New Vanguard for the Environment: Grass-Roots Ecosystem Management as a New Environmental Movement. *Society and Natural Resources*, **13**, 237–259.

Weber, Edward P. 2003. *Bringing Society Back in: Grassroots Ecosystem Management, Accountability, and Sustainable Communities*. Cambridge, MA: The MIT Press.

Weber, Edward P. 2009. Explaining Institutional Change in Tough Cases of Collaboration: "Ideas" in the Blackfoot Watershed. *Public Administration Review*, **69**, 314–327.

Weber, Sindhu, Morris, John C., Krawczyk, Kelly, and Diaz-Kope, Luisa. 2024. Commitment to Place as Motivation for Citizen-Based Watershed Collaboration: The "BIMBY" Effect. *Journal of Public and Nonprofit Affairs*, **10**(2), 222–238.

Williams, Andrew P. 2016. The Development of Collaboration Theory: Typologies and Systems Approaches. In *Advancing Collaboration Theory: Models, Typologies, and Evidence*, edited by John C. Morris and Katrina Miller-Stevens, 14–42. New York: Routledge.

Williams, Christopher M., Merriman, Connie, and Morris, John C. 2016. A Life-Cycle Model of Collaboration. In *Advancing Collaboration Theory: Models, Typologies, and Evidence*, edited by John C. Morris and Katrina Miller-Stevens, 175–196. New York: Routledge.

1

Introducing the Water Governance Puzzle

Collaboration first appeared in the policy and administration literature in 1985 with the publication of an article by Barbara Gray (1985). Gray's argument was straightforward: Government cannot fix all problems – particularly those that are complex, persistent, and resistant to resolution (see Rittel and Webber 1972). She envisioned a setting in which interested parties would coalesce around a common problem, arrive at a shared goal, develop a plan to achieve it, and apply their resources in pursuit of that goal. These actions were framed as a voluntary effort among engaged stakeholders. In essence, to mitigate society's most intractable challenges, actors needed to pool resources and authority beyond traditional command-and-control structures.

Despite the intuitive appeal of such efforts, it remains unclear whether collaboration effectively addresses these types of challenges. Equally important, if it does work, what conditions contribute to its success? Water quality governance offers a particularly revealing case. It exemplifies what Rittel and Weber (1973) define as a wicked problem: A multifaceted, politically contested issue with no single technical solution. Several vexing dimensions of water quality management illustrate this complexity:

1. How actors define water quality and water use matters – as those who enjoy the status quo and the upper hand have the advantage in getting their preferred solution onto the decision agenda as compared to other actors (see Heilman et al. 2024).
2. Water quality has no "stopping rule." In other words, the challenge of securing water quality is never quite resolved as stakeholders can never quite determine when the job is done. In fact, even when standards are set and enforced, they must continually evaluate – as individuals hold differing risk thresholds, preferences, needs, and vulnerabilities. Moreover, new threats emerge. If the work of managing the nation's water quality were to stop, it is not because

stakeholders have solved the problem, but rather, external factors have pushed the issue off the agenda.

3. The solutions to water quality are not binary, dichotomous, or easy. Rather, potential solutions involve a complex set of tradeoffs and values, such as costs, risk levels, equity, vulnerabilities, and so on. Moreover, actors' preferred positions, solutions, and judgments are contingent on their related interests, values, and ideology.
4. Water quality challenges are place-based as well as trans-jurisdictional. For the former, the actors and challenges are unique to that setting. As for the latter, watersheds ignore political and human-drawn boundaries and, as a result, can require extensive intersectoral and intergovernmental cooperation. Here, the specific problems and solutions have important contextual differences that make one-size-fits-all approaches challenging while also mandating coordination across a myriad of actors.
5. Water quality is related to a multitude of additional challenges/problems ranging from injustice and equity to technological and trust.

Thus, as a wicked challenge, water quality and watershed governance are a particularly useful lens for the study of collaboration and, if collaboration works, to achieve desired outcomes. Morris et al. (2013) explain that watershed collaboration offers a local connection (and affinity) for the people who live and work in (and adjacent to) that watershed, a phenomenon they call "BIMBY" (Because It's My Back Yard). These individuals directly experience the costs of impaired waters, and by extension, the benefits of more adequate waters. The extant literature is replete with studies that examine watershed collaboration at the local level and often researchers find evidence of largely successful collaborations (at least, individuals perceive improved outcomes), but it is less clear whether this is a true reflection of the impacts of collaboration or an artifact of the choice of cases. Likewise, while we know much about the "success" factors of watershed collaboration, much less is known about the factors that lead to unsuccessful collaboration (McNamara, Miller-Stevens, and Morris 2020). There are significantly fewer studies that assess outcomes across a large number of collaborations. This gap precludes the elucidation of more generalizable lessons about the factors associated with watershed collaboration and the effects of collaboration in improving water quality. Recognizing this gap, we offer a study that assesses the efficacy of collaboration and its ability to improve water quality.

To do so, we build and engage a large-*n* study that offers a more representative understanding of the dynamics and outcomes of collaborative watershed management. In short, we unearth and tell the story of the relationships among watershed stakeholders and water quality outcomes (both revealed and perceived) in a sample that represents nearly 25% of all watersheds across the United States. Our primary research question investigates whether water quality is higher in watersheds with collaborative partnerships than in those without such arrangements. A secondary research question probes the ecological and social characteristics and types of organizations associated with measurable and perceived improvements in water quality.

1.1 Defining Collaboration

One of the hallmarks of the collaboration literature is the variation in the definition of the term under study. Barbara Gray (1989, 5) provides what is likely the first working definition of collaboration. She suggests that collaboration is

> a process through which parties who see different aspects of a problem can constructively explore their differences and search for solutions that go beyond their own limited vision of what is possible.

This statement, while very broad in scope and coverage, nonetheless provides an initial conception to develop a more specific definition. McNamara (2012, 391) follows a similar line of argument, but adds additional clarification:

> … collaboration is defined as an interaction between participants who work together to pursue complex goals based on shared interests and a collective responsibility for interconnected tasks which cannot be accomplished individually.

Thomson (2001, 83) adds still other elements to her definition collaboration:

> Collaboration is a process in which autonomous actors interact through formal and informal negotiation, jointly creating rules and structures governing their relationships and ways to act or decide on the issues that brought them together; it is a process involving shared norms and mutually beneficial interactions.

More recently, a process reported by Morris and Miller-Stevens (2016) illustrates the challenges when thinking about a definition of collaboration. They report an exercise in which a group of PhD students, participants in a doctoral seminar on collaboration, were tasked with reviewing the extant literature on collaboration and making note of the definitional elements offered in that literature. The students noted over 60 unique elements in definitions offered across a set of only 30 articles, and less than a third of those were common across 2 or more of the reviewed articles (Morris and Miller-Stevens 2016: 5). Lacking a common definition, how do we recognize the phenomenon under study?

The definitional issue is complicated by the fact that collaboration is often conflated with other, similar, forms of interaction. For example, how do we tell the difference between cooperation, coordination, and collaboration? All are common forms of interaction, and all three terms are commonly found in the academic literature. To make matters more complex, all three terms are also very common in practice and are often used interchangeably in both academia and practice. Is there a meaningful distinction, or is this a distinction without a difference? Relatedly, are there performance differences among the interactions?

Williams (2016) and Mayer and Kenter (2016) have addressed this fundamental question. Williams (2016) notes that many frameworks of collaboration in the literature offer thematic or membership characteristics for collaboration, without actually defining how the characteristics differ. He also notes that typologies offer more promise in terms of separating adjacent interactions, but also points out that little work has focused on "rigorous theoretical conceptualization" (Williams 2016, 15). Through a comprehensive review of the literature, Mayer and Kenter (2016) list the most common characteristics of collaboration and arrive at a list of nine characteristics. McNamara (2016) follows a similar methodology but creates a matrix that lists the differences in characteristics across a set of four interactions: cooperation, coordination, mandated collaboration, and voluntary collaboration. Her matrix consists of 10 elements, each of which is uniquely defined across the four interactions. In this sense, McNamara's contribution gives us, arguably, the most comprehensive and theoretically grounded path to distinguish collaboration from other forms of interaction. We can thus define collaboration as a form of interaction by the presence of the 10 elements listed, defined as noted in Table 1.1.

Table 1.1 *Elements of collaboration as a form of interaction. Each element reflects structural or behavioral features that enable collective action across organizations.*

Element	Definition
Design	Shared power arrangements
Formality of the agreement	Informal and formal agreements
Organizational autonomy	Not autonomous; policies to govern the collective arrangement are developed jointly by participants
Key personnel	Not autonomous; policies to govern the collective arrangement are developed jointly by participants
Information sharing	Open and frequent communications through formal and informal channels
Decision-making	Participative decision-making
Resolution of turf issues	Participants work together to resolve conflicts
Resource allocation	Physical and nonphysical resources are pooled by participants in support of collective goals
Systems thinking	System integration does occur to better achieve collective goals
Trust	Trust relationships at all organizational levels is needed as an input to sustain relationships

1.2 A Brief Review of the Watershed Collaboration Literature

As noted previously, the term "collaboration" first entered the scholarly lexicon in a structured manner in 1985 with the publication of Barbara Gray's (1985) seminal article in *Human Relations*. In this article, and in her subsequent book (1989), Gray formulated collaboration as a form of interorganizational interaction with specific characteristics. The motivation for collaboration, she argued, was that intractable problems were beyond the capacities of individual organizations, thus requiring a great effort. Moreover, the mechanism through which collaboration worked was fundamentally different: Rather than a directive or authoritative relationship between entities, or a conflictual relationship, she conceived of collaboration as an interaction in which resources and goals were shared among participants. Another important feature of Gray's formulation was that collaboration was voluntary in nature.

Around the same period of time, other scholars of public administration, public policy, and other disciplines began to focus on network relationships as lens through which to view policy formulation, implementation, and other related activities (Hale 2011). It did not take long for other scholars to note the similarities between collaboration and network theory, and before long a body of literature appeared that treated collaboration as a set of network relationships (see, e.g., Agranoff 2008; 2007; 2006; Keast et al. 2004; Kelley and Morris 2010). Around the same point in time, institutional design scholars also began to note similarities between collaboration and the tenets of public choice theory, leading to yet another conception of collaboration grounded in the idea of self-organizing institutions (see Ostrom 2005; 1999, 1989; Ostrom, Gardner, and Walker 1994; Feiock et al. 2010) formed to address collective action problems (see Scholz et al. 2008). Still another stream of literature conceived of collaboration as a public management function; this literature was largely concerned with providing useful guidance to public managers about how to manage effectively in a collaborative framework (see Agranoff and McGuire 2003; McGuire 2006; Cooper et al. 2006; McNamara 2012).

Emerging concurrently was a particular focus on collaboration as a mechanism to address watershed protection. Although collaboration was applied to many different policy arenas, the focus on, and link to, watershed protection was notable. Lubell et al. (2002) observed this focus, as did Sabatier et al. (2005), Duram and Brown (1999), and Weber (2003). Moore and Koontz (2003) offered a useful typology that distinguished watershed partnerships by the dominant category of members: citizen-based, agency-based, or mixed. Largely based on the work of Moore and Koontz, other scholars noted that agency participation in collaboration tended to vary as well, which led to the development of a distinction between *voluntary* collaboration (in Gray's (1989) sense) and *mandated* collaboration (McNamara 2016; see also Martz et al. 2018). Because public authorities are subject to legal-rational authority, they can be required (mandated) to collaborate by higher political authorities.

As the work noted in the preceding text continued, other work in the literature(s) began to sift through the multitude of theoretical issues surrounding collaboration as an interaction. These issues ranged from the basic conception of collaboration as an interaction (see Williams 2016) to the definitions of the basic elements of collaboration (McNamara and Morris 2012; Diaz-Kope et al. 2014; Bryson et al. 2006; Leach et al. 2002). Other work began to return to Gray's (1989; 1985) initial conception of collaboration as a form of organization (Williams et al. 2016).

1.3 An Urgent as Well as Political Need

The urgency of addressing the nation's water quality is undeniable. Whether it is due to the increasing threats related to climate change (and the associated climatic changes leading to increasing droughts, floods, demands, and pollution), the institutions and actors associated with water quality (see Morris 1996) are confronting a threat-filled, complex and conflict-laden environment. The role of government in the delivery of water quality, however, has evolved slowly over time, and continues to evolve to the present day (Morris 2022). Today, the provision of clean water, and the assurance of adequate water quality, is an essential service provided by government (either directly or via contract). Clean water is not a given, but its importance to life is paramount. In fact, contemporary society relies on water for uses that touch on nearly all facets of human life, from drinking to bathing, cooking, agriculture, transportation, manufacturing and economic production, and recreation. In other words, access to water, especially clean water, is essential to human life.

In the first century of the American republic, and particularly in the territory east of the Rocky Mountains, clean water seemed plentiful for most uses. Water was used as a medium to carry away waste, and the land seemed large enough (and with enough capacity) that the waste did not seem to have a deleterious effect on those downstream. Little attention was given to water, and most water issues were addressed through private action. By the late 1800s, though, water issues began to capture the attention of national policymakers. The Rivers and Harbors Act of 1899 was the first federal legislation to address water quality; the act prohibited ships in port from dumping waste into the harbors (Morris 2022). Of note is that this legislation did not treat water as an environmental issue; rather, the concern was focused on the ability of ships to transit in and out of port safely.

In the early 20th century, as science connected human waste in water to various diseases, some states promulgated laws to prevent its disposal into waterways and encouraged local governments to collect and treat contaminated water. The national government joined this movement for the first time in 1948 with the passage of the Federal Water Pollution Control Act (P.L. 80-845). This legislation was amended and expanded several times over the next 24 years, culminating in the passage of the landmark amendments to the Federal Water Pollution Control Act of 1972, more commonly referred to as the Clean Water Act (Morris 2022). The 1972 legislation was amended several more time over the ensuing 15 years; as of this writing, the Water Quality Act of 1987 stands as the most recent revision of federal water quality policy.

During the signing of the Water Quality Act of 1965, for example, President Lyndon Johnson emphasized the urgent need to stop the polluting of America's waterways. He noted that such activities had turned America's waters from pristine national treasures into dumping grounds for waste. In many ways, President Johnson's remarks signaled a shift from subnational efforts to a more robust federal role to protecting the nation's waterways, as evidenced in remarks made during his signing of the Water Quality Act of 1965:

> The clear, fresh waters that were our national heritage have become dumping grounds for garbage and filth. They poison our fish; they breed disease; they despoil our landscapes.
>
> No one has a right to use America's rivers and America's waterways that belong to all the people as a sewer. The banks of a river may belong to one man or even one industry or one State, but the waters which flow between those banks should belong to all the people. There is no excuse for a river flowing red with blood from slaughterhouses. There is no excuse for papermills pouring tons of sulphuric acid into the lakes and the streams of the people of this country. There is no excuse – and we should call a spade a spade – for chemical companies and oil refineries using our major rivers as pipelines for toxic wastes. There is no excuse for communities to use other people's rivers as a dump for their raw sewage. This sort of carelessness and selfishness simply ought to be stopped; and more, it just must be reversed. And we are going to reverse it. We are going to begin right here in Washington with the Potomac River. Two hundred years ago George Washington used to stand on his lawn down here at Mount Vernon and look on a river that was clean and sweet and pure. In our own century President Theodore Roosevelt used to go swimming in the Potomac. But today the Potomac is a river of decaying sewage and rotten algae. Today all the swimmers are gone; they have been driven from its banks.
>
> *– President Lyndon B. Johnson, 1965 (Johnson 2024)*

In spite of growing governmental involvement, especially at federal regulatory policy, gains in water quality (especially in some watersheds) have been more problematic and uneven. River systems still suffer the ill effects of centuries of industrial activity and poor waste practices; in other parts of the nation watersheds are under threat from diffuse sources of pollution (referred to as nonpoint water pollution). Yet, as compared to the mid twentieth century, many US waters are significantly cleaner, a point made by then-President Bill Clinton. His remarks, quoted below, reflect the continued national-level commitment to water quality:

> To those who say we have nothing more to do to clean up America's waterways, I urge them to come here to Pierce Mill and read the sign. We still have a lot of work to do on this, the most simple necessity of our lives, water. Pierce Mill and this part of Rock Creek Park are very important in the history of our country. Teddy Roosevelt used to come here to walk and to look at the creek, to get a little exercise... we have worked as one people for 25 years – as one people for 25 years – across party lines to make our environment safer and cleaner... Every year the real Clean Water Act cleans more than a billion pounds of toxic pollutants from our water. Every year it keeps 900 million tons of sewage out of our rivers, lakes, and streams. In human terms, it keeps poisons out of your child's evening bath and bedtime glass of water... because of the genuine Clean Water Act, that's on its way to being a dark and distant memory... A big part of the American dream goes way beyond economics and has to do with the preservation of our liberties and the stewardship of our land. This is a part of the American dream.
>
> – *President William J. Clinton, 2024 (Clinton 2024)*

Even with decades of national and state government intervention, funding, and regulation, water quality issues persist in many areas. In fact, a recent national survey of the 20-year water quality needs in the United States put the total figure at more than $630 billion (EPA 2008, 1); much of that figure is needed just to meet existing federal regulatory requirements. President Joe Biden, during a 2023 speech at the Belmont Water Treatment Plant, in Philadelphia, Pennsylvania, pointed to the magnitudes of the water-related challenges ahead:

> I'm here today to talk about an issue affecting Americans across the country: water. You think it's silly, but water – you know, the issue has to do with basic dignity. Every American deserves to be able to turn on their water tap or faucet and be able to drink clean water. [W]e're the richest, most prosperous nation in the world. Water ought to be something that's just guaranteed.
>
> But, unfortunately, that's not the case. Right now, there are an estimated 160,000 service lines carrying water into homes and schools in Pennsylvania, including an estimated 20,000 lines for homes here in the state of – in the city of Philadelphia. Ten million families across the country and hundreds of thousands of schools [and] daycare centers exist [where] their water is pumped through lead pipes, lead valves, and lead fixtures. And, folks, look, the problem is almost in every state, but it's especially bad in older cities in the Midwest and here in the Northeast.
>
> – *President Joseph R. Biden, 2023 (Biden 2023)*

The comments of Presidents Johnson, Clinton, and Biden highlight the uneven progress made in securing adequate water quality. Each also framed clean water as a fundamental human right or part of the great American experiment. Despite progress, what has become increasingly clear to many observers is that government policy intervention and government resources are insufficient to meet the water quality needs of the nation. In short, there are too many waterways and too many problems for government to solve alone. Into this breach new forms of governance have emerged that seek to move beyond command-and-control policy intervention (Ernst 2010) to address water quality issues at the watershed and/or more local level. These new forms of governance seek to accomplish what government has not, and cannot accomplish alone: the protection and restoration of watersheds. These efforts often employ a specific form of governance, known as collaboration. The key question, therefore, is does collaboration work – and if so – what are the outcomes?

1.4 Moving toward Collaboration

From the eighteenth and nineteenth century through the mid twentieth century, water evolved from a taken-for-granted and abundant resource to a central concern of policymakers. Today, as a resource, water is an integral part of conversations centering on environmental sustainability, quality of life, and even the American way of life. This shift has transformed water policy from a fragmented, locally managed issue into a complicated set of water quality and management policies, driven by federal standards. However, federal policy is incomplete and oftentimes inadequate for twenty-first century water quality challenges, which has paved the way for alternatives to traditional regulation, such as collaboration.

US environmental, as well as water quality, policy did not emerge in a vacuum. Klyza and Sousa (2008) write that throughout the past 200 years, US environmental policy evolved over three distinct phases (or layers), identified as economic liberalism, conservation, and environmentalism, as shown in Table 1.2. Importantly, new layers did not replace the previous layer(s) but were laid atop existing policies, values, and institutional structures. The economic liberalism layer emphasized resource development, minimal regulation, and strong property rights – a logic that remains embedded in long-standing agencies like the Army Corps of Engineers. The conservation layer added professional management and preservation of natural areas, while the environmentalism layer expanded legal mandates, public engagement, and decentralization. The net effect, they argue, created (and is creating) a complex and often contradictory labyrinth of environmental laws and institutions that are available to stakeholders.

1.4.1 Golden Era: First Generation Policies

The first Earth Day celebration in 1970 is often considered a pivotal moment for the development of the "green state." Here, strong public opinion in favor of environmental protection, coupled with increased congressional and presidential attention, contributed to a period of frenetic policy activity at the national level.

Table 1.2 *Layers of environmental policy in the United States over time, with dominant values and institutional tendencies characteristic of each period.*

Layer (Time Frame)	Dominant Values
Economic Liberalism (Founding–1890s)	• Develop the nation's natural resources • Sustain a light administrative and regulatory footprint • Value property rights, economic growth, and minimal regulation • Remarkable durability (e.g., Army Corps of Engineers still handles water development)
Conservation (1890–1960s)	• Prioritize scientific and professional management for sustainable yields • Preserve scenic areas, such as national parks
Green State or Golden Era (1960s–Today)	• Promulgate new laws (e.g., Clean Water and Safe Drinking Water Act), expand agency missions • Increase tensions and conflict, especially among extractive groups • Expand roles for citizens, private sector, and nonprofits • Increase tendencies toward decentralization, budget cuts, and devolved responsibilities to states

Source: Klyza and Sousa (2008).

David Webber (2008) suggests that the convergence of congressional and executive policy streams in the late 1950s and 1960s prepared the way for, and gave momentum to, the Earth Day movement. Rachel Carson's (1962) cautionary tale, *Silent Spring*, had a lasting effect and made much of the American public aware of myriad environmental menaces and their likely causes and outcomes. Likewise, the creation of the EPA in 1970 provided, for the first time in American history, a centralized implementation agency for environmental policy (Andrews 2019). The new agency was not only granted implementation authority by Congress, but crucially, regulatory powers as well.

As political forces began to mobilize around environmental issues, an avalanche of national legislative and administrative action occurred between 1970 and 1976. This period saw the passage of landmark laws, including the National Environmental Policy Act (1970), the Clean Air Act (1970), the Federal Water Pollution Control Act – commonly known as the Clean Water Act – (1972) Amendendments to the Federal Water Pollution Control Act (P.L. 92-500), the Noise Control Act (1972), the Coastal Zone Management Act (1972), the Marine Mammal Protection Act (1972), the Federal Environmental Pesticide Control Act (1972), the Endangered Species Act (1973), the Safe Drinking Water Act (1974), the Energy Policy and Conservation Act (1975), the Toxic Substances Control Act (1976), and the

Resource Conservation and Recovery Act (1976), among others. These laws were partly a response to growing public concern over a range of national issues – from civil rights to the Vietnam War (Repetto 2006) – but their enactment was made possible by a rare alignment of political will, policy opportunity, and problem recognition. As Webber (2008) argues, this confluence opened a policy window that punctuated the prevailing policy equilibrium. Repetto (2006) emphasizes the uniqueness of this moment in American environmental history: There was nothing like it before, and there has been nothing like it since.

The institutions that grew out of this "environmental decade" adopted a command-and-control compliance strategy. Here, the base assumptions held that without strict, adversarial, and robust oversight, regulated entities would likely pollute. This approach overcame many collective action problems, such as point-source pollution and resource degradation. Rivers rarely catch fire today, and numerous endangered species have remarkably recovered. For these types of problems, traditional institutions have proven overwhelmingly effective. Other problems, however, have now come to the fore; they require different, softer solutions and more modern approaches (Durant, Fiorino, and O'Leary 2004).

1.4.2 Golden Era: Second Generation Policies

Durant, Fiorino, and O'Leary (2004) summarized first-generation approaches as rule-based models of rigid, adversarial, and reactive command-and-control regulatory regimes. They acknowledge that this regulatory approach successfully addressed many pressing environmental problems in the mid to late twentieth century. However, first-generation approaches, they argue, failed to recognize that many environmental threats are inherently interactive, involve multiple causal pathways, and are cumulative in nature (Durant, Fiorino, and O'Leary 2004, 3). Moreover, first-generation approaches failed to encourage cooperative, holistic, and cost-effective responses to more contemporary problems (Durant, Fiorino, and O'Leary 2004, 4). Such an approach, they contend, is now inadequate in addressing intractable environmental problems such as nonpoint pollution, species protection, biodiversity, and habitat restoration (Agranoff 2012; Durant, Fiorino, and O'Leary 2004, 2–4).

Today's problems, according to Durant and colleagues (2004, 7), necessitate the development and deployment of second-generation regulatory strategies that reconceptualize purpose, reconnect with stakeholders, and redefine administrative rationality. They add that environmental institutions must encourage civic environmentalism, collaboration and deliberative democracy in ways that enhance the legitimacy and quality of decision-making, develops trust, and fosters social capital, or networks of reciprocity. These institutions, as described by Durant and colleagues (2004), must focus on place-based problems and solutions and should be flexible enough to adapt to changing social and natural system conditions.

1.4.3 The Green State Today

The green state is replete with policies that represent each of the layers as well as the deep footprint first-generation enforcement practices. Because the layers grind against one another leading to friction among policies, they often encourage the involvement of multiple institutions and take place in multiple venues. Moreover, the layering of multiple policies and institutions, without the clearing of previous policies and priorities, has led to a contemporary environmental policy landscape where new policies and approaches often conflict with older ones but also provide room for innovation to occur.

1.5 Twenty-First Century Challenges to Water Quality

A range of factors continue to threaten water quality in the US, operating alongside the already complex legal and regulatory environment. On the technical front, pollution detection, testing, and treatment must keep pace with evolving science and emerging risks – including contaminants of emerging concern (CECs). From a governance standpoint, protecting water quality demands an all-hands-on-deck approach – one that engages a broad array of stakeholders, especially in addressing diffuse and unregulated runoff. At the same time, replacing, repairing, and upgrading infrastructure, as well as complying with regulatory mandates, requires substantial upfront and ongoing investment across multiple institutions (see Morris 2022). Industrial activity, energy production, and agriculture all exert constant pressure on water systems. In short, ensuring clean water remains a persistent and formidable challenge.

1.5.1 Acute Water Challenges

Water quality is shaped by a complex interplay of physical, chemical, and biological factors. Contaminants may originate from natural sources – such as minerals and organic matter – or from human activities, including industrial discharges, stormwater and agricultural runoff, and urbanization. Increasingly, water systems must also contend with CECs, a broad category that includes pharmaceuticals and personal care products (PPCPs), microplastics, and per- and polyfluoroalkyl substances (PFAS). These pollutants pose new challenges for monitoring and treatment, often requiring advanced technologies and coordinated regulatory responses. Many of these threats are summarized in Table 1.3.

Table 1.3 *Major types and sources of water quality challenges in the United States. Descriptions reflect EPA language.*

Type	Source	Description
Industry and agricultural	Anthropogenic	Organic solvents, petroleum products, and heavy metals from disposal sites or storage facilities can migrate into aquifers. Pesticides and fertilizers can be carried into lakes and streams by rainfall runoff or snowmelt, or can percolate into aquifers.
Human and animal waste	Anthropogenic and biological	Human wastes from sewage and septic systems can carry harmful microbes into drinking water sources, as can wastes from animal feedlots and wildlife. Major contaminants include *Giardia*, *Cryptosporidium*, and *E. coli*.
Treatment and distribution	Infrastructure and operational	While treatment can remove many contaminants, it can also leave behind byproducts (such as trihalomethanes) that may themselves be harmful. Water can also become contaminated after it enters the distribution system, from a breach in the piping system or from corrosion of plumbing materials made from lead or copper.
Emerging contaminants	Anthropogenic	Contaminants of emerging concern (CECs), including pharmaceuticals and personal care products (PPCPs), are found in waterways and may have ecological or health effects not yet fully understood.

1.5.2 Governance Challenges

Water policymaking in the US can confound and complicate, as it involves numerous institutions, multiple challenges, various uses and goals, a myriad of actors, numerous decision-making points, different levels of influence and access, and often the complexities of federalism, all of which provide inputs into the political system (Heilman et al. 2024). Multiple characteristics and the politics that influence them interact and intersect, and the output is some sort of policy prescription that must satisfy multiple uses. In fact, there are four categories for intended uses for surface waters such as rivers and lakes: human water supply, human recreation, ecological support for fish and wildlife, and economic support for industry, power generation, and navigation (Hoornbeek 2011, 29–30).

The institutional framework for US water policy is best described as a fragmented arrangement of complex legal and administrative structures "based on a wide variety of federal laws, regulations, and historical court rulings, to distribute authority over water between federal, tribal, state, and local governments" (Christian-Smith and Gleick 2012, 23). Evidence of this fragmentation can be seen in the 30 agencies housed in 10 different federal departments with some responsibility over water management – alongside independent agencies, commissions, councils, and other offices (Christian-Smith and Gleick 2012: 28). At the national level, for example, numerous agencies oversee specific attributes of watershed governance, including the Army Corps of Engineers, the Environmental Protection Agency, the Bureau of Reclamation, the Federal Energy Regulatory Commission, the Fish and Wildlife Service, and the Forest Service, as well as overlapping federal statutes such as the Endangered Species Act, the Clean Water Act, and the Clean Air Act. This proliferation of responsibilities is often mirrored at the state and local levels, further fragmenting authority within a single watershed. To complicate matters, hydrographic boundaries rarely align with political ones, demanding coordination both across jurisdictions and between levels of government. Even when agencies share geographic space, they often work at cross-purposes – creating situations in which environmental priorities are subordinated to institutional agendas (Schlager and Blomquist 2008, 190).

Organizations must also confront more traditional challenges related to information asymmetries, boundedly rational people, and boundary problems. The net effect of this set of challenges often means that organizations have difficulty reaching beyond their constituencies to gather and channel resources in efficient and effective ways to address complex problems. Many first-generation responses from command-and-control institutions were effective at meeting early challenges. Yet, because of the nature of second-generation problems – those that reflect the complexity and interconnectedness of policy problems and possible solutions – command-and-control organizations are less effective (Agranoff 2012; Bingham and O'Leary 2008; Rabe 2009). Further complicating water quality governance is a combination of devolutionary processes, uncertainty, rapid technological innovation and change, increasingly scarce resources, and growing organizational interdependencies (Thomson and Perry 2006, 20).

Andrea Gerlak (2006) suggests that challenges of policy fragmentation and ecosystem complexity remain, as does the tension between national policy priorities and local autonomy. Gerlak believes water policy has evolved into a form of pragmatic federalism, consisting of place-based collaborative experiments, emphasizing process rather than division of authority. Multiple competing interests and stakeholders contest today's water policy arena, numerous lawsuits and legal battles, demographic changes, and the lack of unifying policy. This view of pragmatic federalism acknowledges the fragmented approach to watershed protection, and seeks to address the inherent obstacles in reform through a watershed-specific (i.e., case-by-case) approach (Gerlak 2006, 243).

1.5.3 Water Conflicts

The conflation of multiple goals, an increasing number of threats, and governance-related complexities has made conflicts related to water seemingly inevitable. O'Leary et al. (2017) categorized environmental politics and conflict by examining the scale and level in which the conflict is occurring, as shown in Table 1.4. The type of conflict matters in the sense that it constrains the number of potential solutions, the actors involved (or able to address a problem) and the venues available to policymakers or third parties.

Some pollutants come from point sources, flowing from drainpipes or other media into lakes and streams. As such, conflicts are usually associated with a small number of actors. Other pollutants come from nonpoint sources, such as roadway runoff, parking lots, and agricultural fields after a rainstorm. Because nonpoint source (NPS) pollution comes from heterogeneous sources and has diffuse qualities, its mitigation or abatement is understood to be a collective action problem. Moreover, it is up to the states and the U.S. Environmental Protection Agency to determine if NPS pollution has had a detrimental effect on water quality. In this context, "impaired waters" is defined as a determination by a state that a specific segment of waterway is polluted to the extent that it is unable to support the intended uses of that waterway under state water quality (Hoornbeek 2011, 30). National and state governments regulate point-source pollution; nonpoint sources of pollution, however, are not (Aistrup and Morris 2024), so they require a different approach if stakeholders wish to prevent or limit their impact on water quality, riparian habitats, and ecosystem processes.

Table 1.4 *Types and descriptions of environmental conflict, organized by location in the regulatory process or physical system.*

Type of conflict	Description
Upstream conflicts	Involve policies, laws, rules, and standards that state how an environmental issue will be addressed.
Midstream conflicts	Involve administrative issues on how resources (e.g., budgetary and personnel) are managed or regulated.
Downstream conflicts	Involve site-specific conflicts such as inspections, testing, and administrative remedies to ensure regulated entities remain (or return to) a state of compliance.
Point source	Involve specific sites such as a drainpipe.
Nonpoint source	Involve diffuse sources and sites such as a roadway.

Source: Generated by authors, drawing on O'Leary (2017).

1.6 The Collaborative Watershed Approach

Former EPA Administrator Carol M. Browner noted the need for collaboration when she introduced the Watershed Approach in 1996 as "[p]eople working together to protect public health and the environment – community by community, watershed by watershed" (EPA 1996). Approaches like these have led to what Robert Agranoff and Mark McGuire (2003), among others, call collaborative public management (see McNamara 2016). Adapting their definition from Agranoff and McGuire (2003), Bingham and O'Leary (2008, 3) define collaborative public management as

> a concept that describes the process of facilitating and operating in multiorganizational arrangement to solve problems that cannot be solved by single organizations. Collaboration means to co-labor, to achieve common goals, often working across boundaries and in multisector and multi-actor relationships.

Collaborative governance involves various entities at multiple levels of government and the private and nonprofit sectors, but it is also one that necessarily involves the public sector and individual citizens interacting in governance structures (Bingham and O'Leary 2008). Whereas watershed collaboration is not unique in its necessity to deal with complex, uncertain, and wicked problems, it is an arena in which these problems are constants, the actors are many, and collective action is required to overcome them (Scholz and Stiftel 2005).

Mark Imperial (2005, 281) analyzed collaborative watershed programs "to examine how collaboration is used to enhance governance networks where problem-solving capacity is widely dispersed and few organizations accomplish their mission by acting alone." He cautions against the notion that collaboration is "a magical cure for all governance problems" Imperial (2005, 282); acknowledging this, he quotes Bardach (1998, 17):

> We should not be impressed by the idea of collaboration per se. That collaboration is nicer sounding than indifference, conflict, or competition is beside the point. So, too, is the fact that collaboration often makes people feel better than conflict or competition. I do not want to oversell the benefits of interagency collaboration. The political struggle to develop collaborative capacity can be time consuming and divisive. But even if no such struggle were to ensue, the benefits of collaboration are necessarily limited.

The evolution of information technology has further paved the way for collaborative approaches, as it has reduced transaction costs and facilitated integration across local and global boundaries (Agranoff 2007; Bingham and O'Leary 2008; Schlager and Blomquist 2008). Advances in data infrastructure, real-time communication, and shared platforms now allow stakeholders to coordinate across jurisdictions, sectors, and institutional boundaries with unprecedented speed and precision. Tools such as geographic information systems (GIS), cloud-based document management systems, and interactive dashboards enable partners to operate from a shared evidentiary base, reducing asymmetries in information and power. This technological shift does more than improve efficiency; it expands the collaborative table by allowing smaller organizations, under-resourced communities, and geographically dispersed actors to participate meaningfully in governance processes. In this way, information technology functions not just as an enabler of collaboration – it becomes a catalyst for broader inclusion and more adaptive, responsive forms of collective action.

In theory, collaboration is a means to gather stakeholders (including public agencies) together in a single forum to engage in consensus-oriented, collective decision-making (Ansell and Gash 2007, 543). In some instances, however, government agencies do not formally participate but listen to concerns and receive recommendations from interested stakeholders who have formed an advocacy partnership (Koehler and Koontz 2008). In other instances, the government(s) – at various and sometimes multiple levels – becomes a partner by providing the necessary resources to encourage group formation. Collaboration, networks, and collaborative governance have emerged with a distinct focus, even if the terms are not uniformly used or consistent.

Water resource managers, watershed organizations, tribes, state and national agencies, and other stakeholders have moved toward and embraced a watershed approach – designed in part by the EPA – for managing water quality and other natural resource characteristics:

> A *Watershed Approach* is a flexible framework for managing water resource quality and quantity within specified drainage areas, or watersheds. This approach includes stakeholder involvement and management actions supported by sound science and appropriate technology. The *watershed planning process* works within this framework by using a series of cooperative, iterative steps to characterize existing conditions, identify and prioritize problems, define management objectives, develop protection or remediation strategies, and implement and adapt selected actions as necessary (EPA 2008, 2–1, emphasis in original).

This approach requires participation from a diverse group of stakeholders, especially when environmental conditions are extremely impaired – the remediation of which is the approach's primary purpose – and when the impairment comes from diffuse and heterogeneous sources, such as NPS pollution.

The watershed approach is actively encouraged by the EPA and other governmental agencies that deal with water and land management; numerous handbooks have been developed over the past three decades, along with numerous capacity-building educational and grant programs that regional EPA offices often oversee. For the approach to work, stakeholders must come together in some form of collaborative effort to identify problems, develop solutions, implement plans, and hold each other accountable. Of course, they also encounter challenges related to the nature of the resource itself, the various threats to water quality, and the governance challenges that are ever-present in water management.

1.7 Watershed Collaboration: Promise or Peril?

One of the enduring debates within the environmental policy community is whether collaboration is a more effective mechanism to address watershed quality issues than top-down regulatory policy proffered by government. Howard Ernst (2010) casts the debate as one between the "Dark Greens" and the "Light Greens." For Ernst, the "Dark Greens" favor strong, authoritative government regulation accompanied by steep penalties to provide a healthy incentive for compliance. The "Dark Greens" argue that non-regulatory approaches to water quality have been largely ineffective, and that they lack incentives for action, or even participation. On the other hand, the "Light Greens" see the role of government not as regulatory but rather as supportive, and that citizens have an inherent duty to act as stewards of the environment. Ernst (2010) also offers a third approach that focuses on market-based mechanisms to drive environmental change. Known as "Cornucopians," this approach suggests that government intervention is unnecessary, and that free markets, left to their own devices, will eventually solve watershed problems.

In his study of efforts to address watershed issues in the Chesapeake Bay, Ernst (2010) notes that "light green" approaches have generally been unsuccessful at the watershed level. He points to a series of efforts among the states in the Bay watershed to address water quality in the Bay and notes that, in spite of all the efforts, the quality of the water in the Bay continues to deteriorate. Because voluntary action has been unsuccessful, he argues, and the "Cornucopia" approach is unrealistic, the only path forward is a strong regulatory framework to compel action. On the other hand, Morris et al. (2013) argue that the missing element in Ernst's (2010) analysis is scale. If one defines the entire Bay as a single watershed (nearly 64,000 square miles of area spread over six states; see Ernst 2003), then Ernst might be correct. However, if one thinks of the Bay watershed as a series of connected watersheds (there are thousands of rivers and creeks that feed into the Bay), then changing the scale of the perceived problem also changes the nature of the problem – along with the options to solve the problem, and the solutions available. In other words, at a reduced scale, "light green" approaches are not only possible; they may be preferable. And, as Morris et al. (2013) argue, the net effects of small-scale collaborative efforts may have greater impact on the overall watershed than a watershed-wide "dark green" approach. If nothing else, the localized nature of collaborative efforts are more likely to engage citizens to accept new practices that help to address water quality (Diaz-Kope et al. 2024; Morris et al. 2019; Morris et al. 2013; McNamara et al. 2010).

1.8 Conclusion

As the US and nations worldwide confront mounting uncertainty from droughts, floods, shifting climatic patterns, and global climate change, the urgency to ensure the sustainability and adaptability of natural resources has never been greater. Climate change is no longer a distant threat – it is a present reality that demands adaptation from both society and our resource management institutions. Escalating conflicts over water, both domestically and internationally, underscore the critical need for new approaches.

The complexity of water governance challenges outlined in this chapter demonstrate that no single entity or traditional approach can address these issues alone. Instead, effective solutions require collaborative efforts that cross jurisdictional and sectoral boundaries, foster intersectoral partnerships and build inclusive governance. By embracing collaboration, stakeholders can pool knowledge, resources, and perspectives to develop adaptive strategies that enhance the resilience and sustainability of our rivers, lakes, estuaries, and other waterbodies, ensuring their viability for both current and future generations.

References

Agranoff, Robert. 2006. Inside Collaborative Networks: Ten Lessons for Public Managers. *Public Administration Review*, **66**(Special Issue), 56–63.

Agranoff, Robert. 2007. *Managing within Networks: Adding Value to Public Organizations*. Washington, D.C.: Georgetown University Press.

Agranoff, Robert. 2008. Collaboration for Knowledge: Learning from Public Management Networks. In Big Ideas in Collaborative Public Management, edited by Lisa Blomgren Bingham and Rosemary O'Leary, 162–194. Armonk, NY: M.E. Sharpe.

Agranoff, Robert. 2012. *Collaborating to Manage*. Washington, D.C.: Georgetown University Press.

Agranoff, Robert, and McGuire, Michael. 2001. American Federalism and the Search for Models of Management. *Public Administration Review*, **61**(6), 671–681.

Agranoff, Robert, and McGuire, Michael. 2003. *Collaborative Public Management*. Washington, D.C.: Georgetown University Press.

Aistrup, Joseph A., and Morris, John C. 2024. Does Institutional Location Matter? The Effect of Agency Type on Resource Distribution for Nonpoint Water Pollution. *Water Policy*, **26**(10), 1039–1055.

Andrews, Richard N.L. 2019. The Environmental Protection Agency. In *Environmental Policy: New Directions for the Twenty-First Century*, edited by Norman J. Vig and Michael E. Kraft, 168–192. Thousand Oaks, CA: Sage.

Ansell, Chris, and Gash, Alison. 2007. Collaborative Governance in Theory and Practice. *Journal of Public Administration Research and Theory*, **18**, 543–571.

Barham, Elizabeth. 2001. Ecological Boundaries as Community Boundaries. *Society and Natural Resources*, **14**, 181–191.

Biden, Joseph R. 2023 (February 3). *Remarks by President Biden on the Administration's Efforts to Replace Lead Pipes and Provide Clean Drinking Water for All Americans*. The White House.

Bingham, Lisa Blomgren, and O'Leary, Rosemary (eds.). 2008. *Big Ideas in Collaborative Public Management*. Armonk, NY: M.E. Sharpe.

Biswas, Asit K. 2004. Integrated Water Resource Management: A Reassessment. *International Water Resources Association*, **29**(2), 248–256.

Blomquist, William, and Schlager, Edella. 2005. Political Pitfalls of Integrated Watershed Management. *Society and Natural Resources*, **18**, 101–117.

Bryson, John, Crosby, Barbara, and Stone, Melissa M. 2006. The Design and Implementation of Cross-Sector Collaborations: Propositions from the Literature. *Public Administration Review*, **66**, 44–53.

Clinton, William J. 2024. *Remarks on Clean Water Legislation*. The American Presidency Project.

Cooper, Terry, Bryer, Thomas, and Meek, Jack. 2006. Citizen-Centered Collaborative Public Management. *Public Administration Review*, **66**(Special Issue), 76–88.

Diaz-Kope, Luisa, Krawczyk, Kelly, Morris, John C., and Weber, Sindhu. 2024. Commitment to Place as Motivation for Citizen-Based Watershed Collaboration: The "BIMBY" Effect. *Journal of Public and Nonprofit Affairs*, **10**(2), 221–238.

Diaz-Kope, Luisa, Miller-Stevens, Katrina, and Morris, John C. 2014. Collaboration Processes and Institutional Structures: Reexamining the Black Box. *International Journal of Public Administration*, **38**(9), 607–615.

Dietz, Thomas, Ostrom, Elinor, and Stern, Paul C. 2003. The Struggle to Govern the Commons. *Science*, **302**(December), 1907–1912.

Duram, Leslie A., and Brown, Katharin G. 1999. Assessing Public Participation in US Watershed Planning. *Society and Natural Resources*, **12**, 455–467.

Elazar, Daniel J. 1987. *Exploring Federalism*. Tuscaloosa: The University of Alabama Press.

Ernst, Howard R. 2003. *Chesapeake Bay Blues: Science, Politics, and the Struggle to Save the Bay*. Lanham, MD: Rowman and Littlefield.

Ernst, Howard R. 2010. *Fight for the Bay: Why a Dark Green Environmental Awakening is Needed to Save the Chesapeake Bay*. Lanham, MD: Rowman and Littlefield.

Feiock, Richard C., and Scholz, John T. (eds.). 2010. *Self-Organizing Federalism: Collaborative Mechanisms to Mitigate Institutional Collective Action Dilemmas*. New York: Cambridge University Press.

Gray, Barbara. 1985. Conditions Facilitating Interorganizational Collaboration. *Human Relations*, **38**, 911–936.

Gray, Barbara. 1989. *Collaborating: Finding Common Ground for Multiparty Problems*. San Francisco: Jossey-Bass Inc.

Grigg, Neil S. 1999. Integrated Water Resource Management: Who Should Lead, Who Should Pay? *Journal of the American Water Resources Association*, **35**(3), 527–534.

Hale, Kathleen. 2011. *How Information Matters: Networks and Public Policy Innovation*. Washington, D.C.: Georgetown University Press.

Heilman, John G., Johnson, Gerald W., Morris, John C., and O'Toole, Jr., Laurence J. 2024. Water Policy Networks in the United States. In *Networks for Water Policy: A Comparative Perspective*, edited by Hans Bressers, Laurence J. O'Toole and Jeremy Richardson, 80–109. London: Routledge.

Hoornbeek, John A. 2012. *Water Pollution Policies and the American States: Runaway Bureaucracies or Congressional Control?* Albany: State University of New York Press.

Imperial, Mark. 2005. Using Collaboration as a Governance Strategy: Six Lessons from Watershed Management Programs. *Administration and Society*, **37**(3), 281–320.

Johnson, Lyndon B. 2024. *Remarks at the Signing of the Water Quality Act of 1965*. The American Presidency Project.

Keast, Robyn, Mandell, Myrna P., Brown, Kerry, and Woolcock, Geoffrey. 2004. Network Structures: Working Differently and Changing Expectations. *Public Administration Review*, **64**(3), 363–370.

Kelley, Amy P., and Morris, John C. 2010. Keeping the Vision Alive: The Role of Networks in National Memorial Building. *Public Voices*, **11**(1), 89–109.

Koehler, Brandi, and Koontz, Thomas M. 2008. Citizen Participation in Collaborative Watershed Partnerships. *Environmental Management*, **41**, 143–154.

Koontz, Thomas M., and Thomas, Craig. 2006. What Do We Know and Need to Know about the Environmental Outcomes of Collaborative Management? *Public Administration Review*, **66**, 111–121.

Leach, William, Pelkey, N., and Sabatier, Paul. 2002. Stakeholder Partnerships as Collaborative Policymaking: Evaluation Criteria Applied to Watershed Management in California and Washington. *Journal of Policy Analysis and Management*, **21**, 645–670.

Lowe, Bobbie, Ostrom, Elinor, Simon, Carl, and Wilson, James. 2003. Redundancy and Diversity: Do They Influence Optimal Management? In *Navigating Social-Ecological Systems: Building Resilience for Complexity and Change*, edited by Fikret Berkes, Johan Colding and Carl Folke, 83–114. New York: Cambridge University Press.

Lubell, Mark. 2000. Cognitive Conflict and Consensus Building in the National Estuary Program. *American Behavioral Scientist*, **44**(4), 629–648.

Lubell, Mark. 2003. Collaborative Institutions, Belief Systems, and Perceived Policy Effectiveness. *Political Research Quarterly*, **56**(3), 309–323.

Lubell, Mark, Schneider, Mark, Scholz, John T., and Mete, Mihriye. 2002. Watershed Partnerships and the Emergence of Collective Action Institutions. *American Journal of Political Science*, **46**(1), 148–163.

Martz, Robert J., McCarthy, William, and Morris, John C. 2018. The Consent Order as Mandated Collaboration: The Case of the Hampton Roads Sanitation District. *Public Works Management & Policy*, **23**(2), 168–185.

Mayer, Martin K., and Kenter, Robert C. 2016. The Prevailing Elements of Public Sector Collaboration. In *Advancing Collaboration Theory: Models, Typologies, and Evidence*, edited by John C. Morris and Katrina Miller-Stevens, 43–64. New York: Routledge.

McGuire, M. 2006. Collaborative Public Management: Assessing What We Know and How We Know It. *Public Administration Review*, **66**(Special Issue), 33–43.

McNamara, Madeleine, Miller-Stevens, Katrina, and Morris, John C. 2020. Exploring the Determinants of Collaboration Failure. *International Journal of Public Administration*, **43**(1), 49–59.

McNamara, Madeleine W. 2012. Struggling to Untangle the Web of Cooperation, Coordination, and Collaboration: A Framework for Public Managers. *International Journal of Public Administration*, **35**(6), 389–401.

McNamara, Madeleine W. 2016. Collaborative Management and Leadership: A Skill Set for the Entrepreneur. In *Advancing Collaboration Theory: Models, Typologies, and Evidence*, edited by John C. Morris and Katrina Miller-Stevens, 116–132. New York: Routledge.

McNamara, Madeleine W., Leavitt, William M., and Morris, John C. 2010. Multiple Sector Partnerships and the Engagement of Citizens in Social Marketing Campaigns. *Virginia Social Science Journal*, **45**, 1–20.

Moore, E.A., and Koontz, T.M. 2003. Research Note: A Typology of Collaborative Watershed Groups: Citizen-Based, Agency-Based, and Mixed Partnerships. *Society & Natural Resources*, **16**(5), 451–460.

Morris, John C. 1996. Institutional Arrangements in an Age of New Federalism: Public and Private Management in the State Revolving Fund Program. *Public Works Management & Policy*, **1**(2), 145–157.

Morris, John C. 2022. *Clean Water Policy and State Choice: Promise and Performance in the Water Quality Act*. New York: Cambridge University Press.

Morris, John C., and Miller-Stevens, Katrina. 2016. The State of Knowledge in Collaboration. In *Advancing Collaboration Theory: Models, Typologies, and Evidence*, Edited by John C. Morris and Katrina Miller-Stevens, 3–13. New York: Routledge.

Morris, John C., Gibson, William A., Leavitt, William M., and Jones, Shana C. 2013. *The Case for Grassroots Collaboration: Social Capital and Ecosystem Restoration at the Local Level*. Lanham, MD: Lexington Books.

Morris, John C., McNamara, Madeleine W., and Belcher, Amy. 2019. Building Resilience Through Collaboration between Grassroots Citizen Groups and Governments: A Case Study. *Public Works Management & Policy*, **24**(1), 50–62.

Olsen, Mancur. 1971. *The Logic of Collective Action: Public Goods and the Theory of Groups*. Cambridge, MA: Harvard University Press.

Ostrom, E. 1985. A Method of Institutional Analysis. In: Kaufmann, F.X., Majone, G., and Ostrom, V. (eds), *Guidance, Control, and Performance Evaluation in the Public Sector*. Berlin: de Gruyter.

Ostrom, Elinor. 1990. *Governing the Commons: The Evolution of Institutions for Collective Action*. Cambridge, MA: Harvard University Press.

Ostrom, Elinor. 1998. A Behavioral Approach to the Rational Choice Theory of Collective Action. *American Political Science Review*, **92**(1), 1–22.

Ostrom, Elinor. 1999a. Institutional Rational Choice: An Assessment of the Analysis and Development Framework. In: Sabatier, Paul A. (ed.), *Theories of the Policy Process*. Boulder, CO: Westview Press.

Ostrom, Elinor. 2005. *Understanding Institutional Diversity*. Princeton, NJ: Princeton University Press.

Ostrom, Elinor, Gardner, Roy, and Walker, James. 1994. *Rules, Games, and Common-Pool Resources*. Ann Arbor: University of Michigan Press.

Ostrom, Vincent. 1974. *The Intellectual Crisis in American Public Administration*. Tuscaloosa: University of Alabama Press.

Ostrom, Vincent. 1999b. Polycentricity (Part 1). In *Polycentricity and Local Public Economies: Readings from the Political Theory and Policy Analysis Workshop*, 52–74. Ann Arbor: University of Michigan Press.

O'Toole, Jr., Laurence J. 1997. Treating Networks Seriously: Practical and Research-Based Agendas in Public Administration. *Public Administration Review*, **57**(1), 45–52.

Rabe, Barry G. 2009. Governing the Climate from Sacramento. In *Unlocking the Power of Networks: Keys to High-Performance Government*, edited by S. Goldsmith and D. Kettl, 62–94. Washington, D.C.: Georgetown University Press.

Rittel, Horst W. J., and Weber, Melvin. 1973. Dilemmas in a General Theory of Planning. *Policy Sciences*, **4**(June), 155–169.

Sabatier, Paul A., and Jenkins-Smith, Hank. 1993. *Policy Change and Learning: An Advocacy Coalition Approach*. Boulder, CO: Westview Press.

Sabatier, Paul A., Focht, Will, Lubell, Mark, Trachtenberg, Zev, Vedlitz, Arnold, and Matlock, Marty (eds.). 2005. *Swimming Upstream: Collaborative Approaches to Watershed Management*. Cambridge, MA: MIT Press.

Scharpf, Fritz W. 1997. *Games Real Actors Play: Actor-Centered Institutionalism in Policy Research*. Boulder, CO: Westview Press.

Schlager, Edella, and Blomquist, William. 2008. *Embracing Watershed Politics*. Boulder, CO: University Press of Colorado.

Scholz, John T., and Stiftel, Bruce (eds). 2005. *Adaptive Governance and Water Conflict: New Institutions for Collaborative Planning*. Washington, D.C.: RFF Press.

Scholz, John T., Berardo, Ramiro, and Kile, Brad. 2008. Do Networks Solve Collective Action Problems? Credibility, Search, and Collaboration. *Journal of Politics*, **70**(2), 393–406.

Thomson, Ann Marie. 2001. *Collaboration: Meaning and Measurement*. Ph.D. thesis, Indiana University.

Thomson, Anne Marie, and Perry, James L. 2006. Collaboration Processes: Inside the Black Box. *Public Administration Review*, **66**(1), 20–32.

United States Environmental Protection Agency. 2008. *Handbook for Developing Watershed Plans to Restore and Protect Our Waters*. Tech. rept. U.S. Environmental Protection Agency, Washington, D.C.

United States Environmental Protection Agency. 2024. *2022 Clean Watersheds Needs Survey: Report to Congress*. Tech. rept. EPA 832-R-24-002. U.S. Environmental Protection Agency, Washington, D.C.

USGS. 2012. *Water Resources of the United States: Hydrologic Unit Maps*.

Walker, Gregg B., Daniels, Steven E., and Emborg, Jens. 2008. Tackling the Tangle of Environmental Conflict: Complexity, Controversy, and Collaborative Learning. *Emergence: Complexity and Organization*, **10**(4), 17–27.

Weber, Edward P. 2003. *Bringing Society Back In: Grassroots Ecosystem Management, Accountability, and Sustainable Communities*. Cambridge, MA: MIT Press.

Weible, Christopher M. 2006. An Advocacy Coalition Framework Approach to Stakeholder Analysis: Understanding the Policy Context of California Marine Protected Area Policy. *Journal of Public Administration Research and Theory*, **17**(1), 95–117.

Williams, Andrew P. 2016. The Development of Collaboration Theory: Typologies and Systems Approaches. In *Advancing Collaboration Theory: Models, Typologies, and Evidence*, edited by John C. Morris and Katrina Miller-Stevens, 14–42. New York: Routledge.

2

Tracing the Development of US Water Quality Policy

Water quality governance in the United States has evolved from a locally centric and piecemeal endeavor to one with a strong federal component, albeit still fragmented in character. This transformation, we suggest, is reflective of clean water's role in promoting public health, ensuring environmental sustainability, and supporting economic development. We begin Chapter 2 with an overview of federal water quality policy by tracking its historical roots and identifying key legislative moments that have subsequently shaped contemporary watershed collaboration efforts. Building on this historical and legal foundation, we then shift to the concepts, definitions, scope, and scales of contemporary watershed governance. We examine how collaborative approaches have emerged as critical tools for managing water resources in a manner more in line with second-generation regulatory approaches. By focusing on watersheds management in practice, this chapter highlights the importance of collaborative governance structures and relationships that are inherent to twenty-first century water quality management and delivery.

2.1 A Brief History of Water Quality Legislation at the Federal Level

The Clean Water Act (CWA) of 1972 (P.L. 92-500), more formally known as the 1972 amendments to the Federal Water Pollution Control Act, stands as the ultimate expression of national water quality policy. While the CWA substantially changed the federal role in water quality, the CWA is more appropriately viewed as a nonincremental change in what was largely an incremental development of federal clean water policy over a span of several decades (Morris 2022). Indeed, the history of the federal role in water quality can be traced to the Refuse Act of 1899 (see Freeman 1990; Eames 1970). This early legislation was arguably more concerned with safety of navigation in rivers and harbors; the act prohibited ships from dumping ballast and dunnage into congested waterways. This material posed a danger to ships and limited free navigation, and was a growing problem in several busy ports. The problem was clearly defined as a transportation, rather than an environmental, problem. However, the Refuse Act contributed one important policy element that would form the basis for all later national policy: For the first time, the act established federal control over navigable waters (Stern and Mazze 1974). The establishment of federal jurisdiction over waterways was critical to later efforts to address water pollution from a public health perspective and, ultimately, as an environmental policy issue.

Following the turn of the century most of the policy efforts connected to water quality were state-level efforts, with more populous cities generally leading efforts to treat municipal sewage. As scientists began to better understand the links between polluted water and disease, efforts grew to curtail the dumping of untreated sewage into the nation's waterways. One of the first federal actions was the Public Health Act of 1912, which authorized the US Public Health Service to conduct water quality research in navigable (i.e., federal) waters (Davies and Davies 1975). As part of the New Deal legislative efforts in the 1930s there were several proposals to fund the construction of municipal wastewater plants as work projects. Similar efforts were made during this period to try to establish federal water quality standards for the first time, but both efforts fell short in Congress or failed to receive the support of the Roosevelt administration (see Morris 2022).

In 1948 a relatively young House of Representatives subcommittee chair, John Blatnik (D-MN), saw an opportunity to expand the purview of his subcommittee (Rivers and Harbors), and to create a program to distribute resources to support water quality efforts (Milazzo 2006). Blatnik proposed a program that would provide grants to smaller municipalities to fund wastewater treatment works. His initial proposal capped the community size at 50,000, but after realizing that such a cap would prevent any communities in his district from receiving grants, the cap was quickly raised to 100,000 (Milazzo 2006). The result of these efforts was the Federal Water Pollution Control Act of 1948. The act itself was modest: It encouraged states to work together to enact pollution requirements for navigable waters; it provided some research funding for water quality; it created the Federal Water Pollution Control Advisory Board; and it authorized a small grant program to provide construction funds for wastewater treatment facilities. The focus on navigable waters is again noteworthy; this legislation built on the federal authority defined in the Refuse Act nearly 50 years earlier. It is notable that Congress (and succeeding Congresses) chose not to appropriate funds for the construction program (Morris 2022), which meant that the act had little real policy impact. However, it did set the stage for future legislation and an expansion of federal authority in this policy realm. The act was reauthorized in 1956 with an expanded construction grants program; in this instance Congress did begin to appropriate funds for the construction grants, a pattern that has continued, nearly without interruption, to the present day.

Over the course of the next 15 years this basic legislation was reauthorized and updated; each update expanded incrementally the federal role in water quality. For example, in 1961 the amendments allowed states to sue other states over interstate water pollution; the 1965 and 1966 legislation shifted the focus of water quality policy from a public health focus to an environmental focus; and the 1970 legislation expanded the purview of water pollution to include oil and hazardous discharges from ships at sea, and moved responsibility for water quality into the newly created Environmental Protection Agency (EPA). In each of these cases, Congress generally chose to expand federal funding support for wastewater treatment plant construction (Morris 2022).

2.1.1 Congress and Point-Source Pollution

The CWA is a cornerstone environmental policy that regulates many of the waters of the US Passed in 1972, during the peak of the environmental golden age, lawmakers sought to restore and protect the environmental integrity of the nation's water supply. To do so, the CWA established a regulatory framework that enabled the newly formed EPA to limit pollutant discharges into waterways, promulgate and enforce water quality standards, as well as providing funding for treatment facilities. This regulatory program, known as the National Pollution Discharge Elimination System (NPDES), was closely modeled on the regulatory framework enacted in the 1970 Clean Air Act (Davies and Davies 1975). The NPDES is explicitly designed to address point-source pollution, and to require states to monitor, report, and enforce permit requirements. The Act also includes provisions for preventing oil spills and requires states to develop and enforce water quality criteria for all navigable waters. Importantly, the act also develops a watershed-based approach to water quality.

The NPDES program is based on a model of shared responsibility for the regulation of point-source water pollution. Under the CWA, states develop standards and limits for pollution discharges, which in turn must be approved by EPA. Upon approval, states can then choose to assume responsibility for the permitting, monitoring, and enforcement of the clean water standards, a process known as primacy (Crotty 1987; Morris 2022). If a state does not accept primacy, then the state cedes responsibility for the NPDES to the national government and EPA. As of this writing, all but four states (New Mexico, New Hampshire, Idaho, and Massachusetts) have assumed primacy for wastewater regulation. Once a state assumes primacy, the state is responsible for reviewing and issuing permits, inspections, reporting, monitoring, and enforcement. The national government only intervenes when states are unable or unwilling to enforce the standards.

The CWA also included language intended to set a standard for the nation's waters: The waters were to be "fishable and swimmable" by 1983 (Poe 1995). This language was itself somewhat controversial, and there was much debate in both the House and the Senate about goals for water quality. Some favored a "zero-discharge" goal, while others argued that the technology to achieve "zero-discharge" was prohibitively expensive (Milazzo 2006). The final version of the law required states to employ "the best practicable technology available" (Morris 2022, 78) to achieve their water quality goals, but "fishable and swimmable" remained undefined. It is worth noting that, in subsequent amendments to the CWA, the deadline for achieving "fishable and swimmable" waters was pushed back, until it was finally abandoned completely in the 1987 legislation (Morris 2022).

2.1.2 Congress and Nonpoint Source Pollution

Following passage of the 1972 Clean Water Act, Congress continued to make incremental adjustments to federal water quality policy. Subsequent amendments defined additional pollutants to be regulated and adjusted the funding formula applied to the construction grants program. Notably, what the several amendments did not address was the federal role in nonpoint water pollution, which remained unregulated. Moreover, there was little pressure on EPA or states to implement the requirements of the 1972 law related to nonpoint water pollution.

In 1986 Congress once again undertook a process to reauthorize the 1972 legislation. One of the main points of interest in this effort was to revise the funding mechanism for federal grants for wastewater treatment construction. The result was a move away from a categorical grant program to a state-administered revolving loan fund program. The new program, known as the Clean Water State Revolving Fund (CWSRF) program, was designed to remove overt federal control over wastewater funding decisions and return control to states (Morris 2022). Known as the Water Quality Act (WQA), the act represented another nonincremental shift in federal water quality policy (Morris 2022). However, the act left in place most of the structure of the CWA; although it also provided more financial support for states to address nonpoint water pollution issues, Congress once again declined an opportunity to regulate nonpoint pollution. Despite the lack of additional regulation then-President Reagan vetoed the legislation in late 1986; Congress reintroduced and passed the bill in early 1987. Reagan once again vetoed the bill, and the veto was quickly overridden by a supermajority in both chambers of Congress (Morris 2022). The 1987 legislation marks the most recent effort to address the federal role in water quality; subsequent legislative activity has made very small changes around program operation (e.g., allow states to comingle wastewater and drinking water revolving funds; a requirement to use US-produced steel in CWSRF-funded projects). While there has been tremendous improvement relative to water quality, significant gaps exist in efforts to protect watersheds from the cumulative impacts, for example, of nonpoint sources and other emerging threats outlined in Chapter 1.

Unlike point-source pollution, Congress explicitly chose not to regulate nonpoint water pollution. This issue was the source of some debate during the development of the 1972 CWA, and again during the 1987 amendments (the WQA), but pressure from farming interests and others effectively removed any possibility of nonpoint regulation.[1] In place of a regulatory framework, Congress developed what is, in effect, a voluntary program known as the Total Maximum Daily Load (TMDL) program. The intent of the TMDL program is for states to determine how much of a given pollutant a segment of a waterway can absorb and still meet water quality standards. If the amount of a pollutant exceeds that level, then a TMDL agreement can be developed to implement voluntary programs to reach the desired standard. The TMDL itself is both an agreement and a water quality standard, in that it sets a goal, and stakeholders agree to implement changes (often called best management practices or BMPs) to achieve the desired goal within an agreed-upon time frame.

2.1.2.1 Key Provisions

Two key provisions of the CWA, Sections 319 and 303, are at the heart of state efforts to achieve, maintain, or protect water quality. Section 319 serves as the legal foundation for addressing nonpoint source water pollution, whereas Section 303 outlines the process for establishing state-specific water quality standards. In 1996, EPA also began a comprehensive evaluation of the TMDL program due to states' recalcitrance or inability to develop TMDLs. At the same time, environmental groups filed lawsuits arguing that the EPA had failed to adequately enforce the CWA, especially in overseeing TMDL progress in states. These sections and related developments serve as the foundation for contemporary water quality management, which we turn to next.

[1] It is worth noting that the possible regulation of nonpoint pollution featured prominently in Richard Nixon's veto of the CWA in 1972, and again in Reagan's vetoes of the WQA in 1986 and 1987. Both presidents used very similar arguments; in effect, they argued that any provisions that addressed nonpoint pollution were, in fact, "stealth" regulatory systems that would have deleterious effects on agriculture in the United States (see the veto messages reprinted in AISWPCA 2004). It is also worth noting that EPA did little to enforce the TMDL requirements of the law until a series of lawsuits brought by environmental groups in the 1990s and 2000s compelled EPA to pressure states to develop TMDLs (see Houck 2002a; Morris 2022).

2.1.2.2 Section 319 in Practice

Section 319 provided the EPA with the legal foundation to address nonpoint water pollution and, as Mayer et al. (2023) write, is generally considered the starting point for state TMDL development. The Section itself mandates that state governors "prepare and submit to the Administrator [of EPA] for approval a management program which such State proposes to implement… for controlling pollution added from nonpoint sources to the navigable waters within the State and improving the quality of such waters" (P.L. 100–4, Sec 319 (b)(1)). In other words, the Section stipulates the following:

1. States must identify the waters that, without intervention, are unlikely to meet water quality standards;
2. States must identify the major pollutants in those waters; and
3. States must identify the best practices available to reduce pollution in these waterways.

Mayer et al. (2023) also note that Section 319 authorizes grant funds to states to assist with assessment and planning. However, as Mayer et al. (2023, 3) note, "it was not until environmental groups began to file suit in the mid and late 1990s against both states and the EPA to force compliance with Section 319 that states began to develop plans" (see also Houck 2002b).

2.1.2.3 Section 303 in Practice

Based on Section 319, Congress established a regulatory requirement on the states to establish water quality standards (WQS) that are applicable to their waters. The law requires states to obtain the EPA's approval of those standards. Within the law, Congress also provided clarity and direction regarding those standards. For the former, CWA section 303(c)(2)(A) stipulates that state policymakers:

> …specify appropriate designated uses of the waters, and water quality criteria to protect those uses. Such criteria must be based on sound scientific rationale, must contain sufficient parameters to protect the designated use, must support the most sensitive use where multiple use designations apply, and may be expressed in either narrative or numeric form.

In addition, 40 CFR 131.10(b) provides that

> [i]n designating uses of a water body and the appropriate criteria for those uses, the state shall take into consideration the water quality standards of downstream waters and ensure that its water quality standards provide for the attainment and maintenance of the water quality standards of downstream waters.

For the latter, the CWA further adds:

> WQS shall be established taking into consideration their use and value for public water supplies, propagation of fish and wildlife, recreational purposes, and agricultural, industrial, and other purposes, and also taking into consideration their use and value for navigation.

Whereas CWA section 303(c) authorizes states to establish WQS initially,

> Congress expressly gave the EPA the responsibility to review state WQS, and to disapprove them and promulgate Federal standards if state standards do not meet the applicable requirements of the Act. (Source: EPA 2024)

Section 303(d) of the CWA outlines the TMDL creation process. It requires states to identify "waters within their borders for which technology-based regulatory controls, i.e., point source pollution are insufficient to ensure in-stream compliance with water quality standards" (Hoornbeek 2012, 422). The defined uses generally determine water quality standards for a specific water body. If applicable, Section 303(d) also requires that the states establish a total maximum daily amount of a pollutant that "a water body or stream segment can receive and still comply with water quality standards" (Hoornbeek 2012, 422), that is, to meet designated use(s). Because the federal and state governments regulate water quality but not nonpoint source (NPS) pollution, this is an essential tool by which to identify and mitigate these pollutants and their impacts on water quality and riparian habitats.

As noted previously, the states establish their own water quality standards, which are reviewed and updated every three years. State standards may exceed the federal standards but cannot fall below those federal standards. A cursory outline of the process is included here:

1. States identify and designate intended uses for bodies of water (e.g., drinking water, irrigation, and recreation).
2. States develop and adopt criteria that describe necessary chemical, physical, and biological conditions that are necessary to protect identified uses. This can include characteristics such as nitrogen levels and concentrations, stream structure, and aquatic life health, respectively.
3. States submit a list of impaired waters to EPA annually.
4. The EPA approves or rejects state TMDL standards and establishes recommended national criteria (if rejected, the EPA develops its own standard within 30 days). State standards "should be comprehensive and strong enough to achieve the Clean Water Act's goal of chemical, physical, and biological integrity" for the state's waters. It is not uncommon for water bodies to be largely free of chemical pollutants yet still vulnerable to other stressors, including stressors affecting habitat, stream morphology (structure), and streamflow, which are critical to supporting aquatic life.
5. After setting water quality standards, states monitor and assess water quality to determine the degree to which the standards are being met. Using existing and readily available water quality-related data and other information, states identify impaired waters that do not meet applicable water quality standards and need TMDLs. EPA's regulations require that states report biennially to

EPA on the quality of their waters, and states generally fulfill this requirement by submitting integrated reports to EPA.

The State of Florida, for example, identifies six uses, which are arranged in order of degree of protection required:

- Class I – Potable Water Supplies. Fourteen general areas throughout the state, including impoundments and associated tributaries, certain lakes, rivers, or portions of rivers, used as a source of potable water.
- Class II – Shellfish Propagation or Harvesting. Generally coastal waters where shellfish harvesting occurs.
- Class III – Fish Consumption, Recreation, Propagation and Maintenance of a Healthy, Well-Balanced Population of Fish and Wildlife. The surface waters of the state are Class III unless described in rule 62-302.400, F.A.C.
- Class III-Limited – Fish Consumption; Recreation or Limited Recreation; and/or Propagation and Maintenance of a Limited Population of Fish and Wildlife. This classification is restricted to waters with human-induced physical or habitat conditions that, because of those conditions, have limited aquatic life support and habitat that prevent attainment of Class III uses.
- Class IV – Agricultural Water Supplies. Generally located in agriculture areas around Lake Okeechobee.
- Class V – Navigation, Utility and Industrial Use. Currently, there are not any designated Class V bodies of water. The Fenholloway River was reclassified as Class III in 1998.

The State's Department of Environmental Protection then determines the body of water's impairment status (Florida Department of Environmental Protection, 2024). Based on state criteria, a body of water is considered impaired if its designated use or uses is (potable, fishable/swimmable, shellfish harvest, etc.) is adversely affected. Addressing impaired waters and/or preventing pristine waters from becoming impaired can involve a variety of strategies as no watersheds (and the sources of their impairment) are identical. We present many of the common approaches in Table 2.1.

2.1.3 CWA Implementation

The EPA began prioritizing its collaborative watershed approach in the mid-1990s; following their lead, the Forest Service and 18 federal agencies that manage land have "adopted ecosystem management approaches that featured collaboration as a central tenet" (Koontz and Thomas 2006, 112). Through the early 2000s, the federal interior and agriculture departments also adopted watershed-based approaches

Table 2.1 *Types of protection strategies used to preserve or sustain water quality in different regulatory contexts.*

Strategy	Description
Protecting waters of higher quality	Creation/application of a protection plan for waters with a higher quality than traditional impairment metrics to ensure that these waters remain protected.
Programmatic protection	Creation/application of a TMDL for a waterbody that is not currently impaired or lacks sufficient information to determine its impairment status. This is more of a proactive approach.
Protection from impairment	Creation/application of a preventative plan or approach for waters that are trending toward impairment but are not officially impaired.
Legacy protection	Creation/application of a legacy protection plan to waterbodies that have been successfully restored and meet water quality standards.

Source: Environmental Law Institute. www.eli.org/freshwater-ocean/compendium-water-quality-restoration-approaches.

that encouraged collaboration with federal agencies, state and local governments, and interested stakeholders. Efforts to prioritize collaboration were also motivated by several administrative factors. First, stakeholders became increasingly dissatisfied with traditional regulatory approaches that failed to address NPS pollution and habitat protection. Second, water systems began to adhere to water quality guidelines outlined in Sections 319 and 303 (CWA), the enforcement of which began in earnest in the mid-1990s (Hoornbeek et al. 2008).

Today, many states share a familiar implementation story related to the CWA. An event happens, such as a lawsuit and eventual settlement or a court order, that demands the development of a TMDL for an impaired body of water. These actions, according to the Environmental Law Institute, may necessitate the development of new programs, policies, and efforts (such as monitoring) – to meet the terms of the consent decree settlement (often in terms of the production of *X* TMDLs by *Y* date). In fact, even states not subject to the settlement or decree may initiate a TMDL following a settlement.

A vicious cycle soon emerged. On one hand, the focus on satisfying consent decrees and/or court orders led to the development of numerous TMDLs; however, this development was not matched with resources to support implementation efforts designed to achieve those standards. These needed resources were not insignificant and presented (and still do) a substantial challenge to many states, which could not or would not fund all needed restoration or protection plans. In fact, in 2022, "the total nationwide water infrastructure needs topped

Table 2.2 *Common restoration activities used to improve water quality. These range from structural improvements to regulatory strategies.*

Restoration activity	Description
Agricultural BMPs	Implementing Best Management Practices (BMPs) to reduce pollution from farming activities.
Other BMPs	Targeting issues related to roads, urban runoff, and mining to address sources of contamination affecting water quality.
Dams/Culverts	Removing or improving dams and culverts to restore natural water flow and protect aquatic ecosystems.
Waste/Sewage collection	Improving waste and sewage systems to reduce pollution from human activities.
Wildlife/Pet management	Managing wildlife and pets to mitigate their impact on water quality.
Source removal	Directly removing the source of impairment, whether a contaminant or a physical barrier.
Riparian area restoration	Restoring natural vegetation along water bodies to stabilize banks and improve habitat.
Environmental remediation	Cleaning up contaminated sites to restore water quality.
Regulatory authority	Exercising regulatory authority to enforce laws protecting water quality.
New laws	Introducing new legislation to address emerging water quality issues.
Incentives	Offering incentives to encourage compliance and adoption of BMPs.
Technical assistance/Education	Providing stakeholders with knowledge and tools for effective restoration.
Collaboration	Ensuring coordinated and comprehensive efforts through collaboration among various entities.

Source: Environmental Law Institute (n.d.).

$630 billion for the period between January 1, 2022, and December 31, 2041 (Mills and Oberthur 2024). On the other hand, other state environmental and water-related needs have continued to grow, pushing many environmental budgets to their limits. In response to limited (or nonexistent) state support, many smaller collaborations have formed at the local level, creating formal and informal policy networks often designed to address a particular issue(s), such as riparian habitat restoration, dam removal, legacy pollution, and NPS pollution, among other environmental concerns. A selection of these restoration activities are presented in Table 2.2.

Finding the solutions to these intractable water quality problems requires "detailed knowledge of local conditions and the coordination of multiple agencies" (Sabatier et al. 2005, 5), creating an opportunity for collaboration. John Hoornbeek and colleagues (2012) acknowledge that the emphasis on collaborative approaches began when the EPA began to enforce TMDL provisions in Section 303(d) of the CWA. Prior to this time, many of the regulations for setting TMDLs were largely ignored by the EPA and the states (for approximately 20 years after the enactment of the CWA). Finally, the federal government has also offered financial incentives to engage in watershed collaboration. For example, in 2004, the EPA Region 5 field office alone awarded $42 million in Clean Water Act (CWA) Section 319 grant funding to "local watershed organizations to develop management plans for their watersheds, emphasizing collaborative, community-based management approaches that use a watershed scale" (Koontz and Thomas 2006, 112).

There are additional reasons that today's EPA is generally seen as amenable to voluntary watershed collaboration.[2] Centered around the language of mission, vision, and values, the EPA notes that watersheds have diverse needs, conditions, and priorities. Because of this diversity, if the agency seeks to prevent pollution, to achieve environmental improvements, and to meet community goals it must pursue a collaborative approach informed via key values: partnerships, geography, and sound management. By engaging a myriad of stakeholders including local, regional, state, tribal, and federal agencies, land and water conservation districts, public and private interest groups, industries, academic institutions, landowners, and concerned citizens, the EPA ensures that the perspectives of those most impacted and whose input is most needed to restore or protect the watershed is informing its decision-making. Moreover, it means that watershed efforts are considered alongside other pressing needs and priorities. Additionally, its focus on specific regions allows for targeted restoration and protection strategies. Finally, by relying on science and data, the varying stakeholders can meaningfully participate in a process that spans the policy process from initial assessment to evaluation (EPA 1996, 2022).

2.2 Watershed Collaboration: What Is Known and What Is Not

Recall that technology-based controls are first-generation responses to pollution – command-and-control approaches. This approach is designed to mitigate point-source pollutants. However, these types of regulatory controls are not ideal for nonpoint sources. And, given the continued challenges associated with NPS pollution, there is a need for a different and more collaborative approach. The question then becomes – why? We turn to several reasons, ranging from the nature of watersheds themselves to the challenges of enforcement and the presence of

[2] These efforts are largely voluntary, but there are examples of required (mandated) collaboration (see McNamara 2016). We focus here on voluntary efforts.

decentralized institutions, that help to explain the possible need for collaboration in the complex arena of watershed governance.

Watersheds vary in their geographic size and location. They are often defined by a particular geographic or hydrographic feature (EPA 2008), and they can overlay predominantly urban or rural settings, or both. The waterbodies contained within watersheds may have a primary purpose of providing water to agricultural interests, or they may store water for small and large communities of people. And, of course, there are numerous uses. The EPA has directed the states to define designated uses for all waterbodies within their jurisdictions. Numerous waterbodies have multiple uses, including recreation, fish and aquatic species propagation, fish harvesting, aesthetic purposes, and navigation, among other uses (EPA 2008).

Watersheds are also nested within larger watersheds, which are nested in still larger watersheds. Agencies and institutions manage watersheds at various scales: local, regional, and super-regional (Blomquist 2009). Moreover, the watershed is affected at these scales, as biotic processes or an effluent or pollutant discharge can have a local or system-wide impact. For example, localized nutrient runoff combined with warm water temperatures may cause algae to bloom in a small segment of a watershed, having only a localized impact; on the other hand, a chemical spill from a mining operation can flow downstream in a serial fashion through several sub-watersheds, potentially flowing to the end of the river system. The agencies that manage, regulate, and develop policies and BMPs for watersheds similarly exist at various scales. For example, the EPA is headquartered in Washington, D.C., with several regional offices distributed geographically; the national parks and forest services are similarly structured.

Furthermore, nearly every state has one or more departments or agencies charged with environmental and natural resources protection, working with the EPA in some capacity. And, of course, there are numerous states, county, municipal, special district, and other agencies located within and across watershed boundaries at whatever scale they are examined. This is why we cannot think about watershed management without considering federalism and intergovernmental relations, and it is why collaboration or collaborative efforts are deemed valuable.

One of the most effective ways to manage a watershed is through a large-scale, integrated management institution (e.g., Ernst 2010; 2003; Grigg 1999) – an institution that could theoretically be capable of managing a region as large as the entire Mississippi River drainage basin. Others do not see this as physically or politically feasible (Biswas 2004; Blomquist and Schlager 2005; Morris et al. 2013). Due to the politics of boundary drawing, decision-making, and accountability, these integrated approaches are challenging and problematic for large geographies. Thus, the most effective way to manage a watershed, according to Blomquist and Schlager

(2005, 113), is through a collaborative approach that involves local stakeholders, providing a way for communities to assert contested claims for inclusion, articulate and protect their values and interests during decisions about the watershed, and invest those decision-making arrangements with the mechanisms for accountability and change. This type of arrangement connects aspects of large-scale national goals with the knowledge and experiences of those who live in and near the watershed (Morris et al. 2013). These arrangements have now become commonplace across the United States – and, again, encouraged by the EPA and numerous other federal departments and agencies charged with land management (Hoornbeek 2012; Sabatier et al. 2005), reflecting the EPA's Watershed Approach (EPA 2023; see also Wayland 2012).

Although the institutionally rich environment, as described, is a possibility, what often occurs is a fragmented arrangement of institutions working either at similar goals or at cross-purposes, adding further complexity and uncertainty to the environment. This fragmentation of authority, goals, and values, among other things, can either push groups apart or bring them together to manage collaboratively and pool resources. Examining this fragmented nature of watershed governance, Mark Imperial (2005, 283) highlights aspects that often make collaboration desirable:

> The policies and programs governing watersheds are specialized by medium (e.g., air, water, soil, land use, etc.), geographic location (e.g., wetlands, coastal zone, tidal waters, agricultural land, forest land, etc.) statute, or function (e.g., permitting enforcement, educating the public, installing best management practices, issuing grants, etc.). The corresponding institutional fragmentation limits any organization's ability to accomplish its mission by acting alone.

So, within each watershed, we can find a institutional environment that deals with these various attributes, as well as commensurate uncertainties and complexities, which are the very nature of both social and ecological systems. Traditional command-and-control institutions have done well to ensure certain parts of the system are efficiently protected from some collective action problems. Other, more difficult collective action problems, such as NPS pollution and habitat degradation, require different institutions and often different institutional actors to deal with a watershed's heterogeneous characteristics in a particular place (EPA 2022).

2.2.1 Watersheds in Practice

Since 1996, watershed partnerships have emerged to mitigate heterogeneous NPS pollution; others have formed to overcome habitat degradation, increase biodiversity, and structure decision-making about the natural resource in ways acceptable to stakeholders within the watershed. The United States Geological Survey (USGS) has classified watersheds into successively smaller hydrologic units at various scales, identified by an eight-digit code, which "is a geographic area representing part of all of a surface drainage basin, a combination of drainage basins, or a distinct hydrologic feature" (USGS 2012). Across the 50 states, there are 2,256 watersheds at this classification level.

As noted, three issues permeate watershed collaborations: boundary definition, decision-making rules and structures, and accountability (Schlager and Blomquist 2008). The political choices confronted, therefore, are structuring governing institutions and organizations, determining who participates in decision-making, deciding how decisions are made (e.g., consensus or majority), and establishing how decision-makers are held accountable. The "[a]rguments over boundaries and inclusion often hinge on the outcomes advocates are trying to promote; they are political arguments" (Schlager and Blomquist 2008, 26). Biophysical and organizational boundaries can change over time, as can values and socioeconomic landscapes:

> Neither defining communities of interest broadly nor giving pride of place to local, geographic communities guarantees that a particular set of values will be pursued consistently over time as the watershed setting and its context change. (Schlager and Blomquist 2008, 63)

While natural features, for example, a lake, river, or stream can sometimes form jurisdictional boundaries, and they often do not match neatly with the political or administrative boundaries drawn by humans. This makes it difficult to overcome collective action problems in natural resource settings. And it also makes difficult the articulation and acceptance of decision-making structures that determine who is accountable to whom.

Schlager and Blomquist (2008) make us keenly aware of the importance of recognizing watersheds' physical and institutional landscapes. The physical landscape captures the geographic, topographic, hydrographic, and hydrologic features, the water resources, the various flora and fauna and their habitats, and the structures and habitats humans have established in the watershed. The institutional landscape of a watershed is often a "complex but largely invisible terrain of rules and organizations that govern and affect human choices about the making of decisions, the use of resources, and the relationships of people to nature and to one another" (Schlager and Blomquist 2008, 1). In Elinor Ostrom's (2005, 17) terms, these rules form the basis for an action situation in each setting or environment. A series of physical, social, and institutional rules structure watersheds and many other natural resources. The action situation is where these rules impact those participating in governance and decision-making.

The terrain of rules is also complex, as formal institutional constraints from multiple levels of government limit, prescribe, proscribe, and enjoin actions. At the national level, the EPA, the U.S. Forest Service, the U.S. National Park

Service, the U.S. Fish and Wildlife Service, the Bureau of Reclamation, and other agencies from the Departments of the Interior, Agriculture, Defense, and Homeland Security all have a stake in managing watersheds across the country. Many of these responsibilities are mirrored, supplemented, or even duplicated at the state and local levels, where agencies deliver public goods and regulatory oversight within watershed systems. While many still perform traditional distributional and enforcement functions, several have evolved into "conductive agencies," working across organizational boundaries to foster collaboration among public, private, and nonprofit actors in pursuit of shared goals (Agranoff 2012).

Watersheds, like ecosystems, are complex adaptive systems. To put a finer point on it, defining an ecosystem is difficult for numerous reasons, so watersheds are often used as "ecosystem proxies" (Barham 2001, 183). Elizabeth Barham (2001, 183) writes:

> Throughout history we have organized our patterns of habitation and food production around water. The placement of villages, towns, and cities, traditional structures of land control and distribution, and indeed the rise of industrial production have all been influenced by the availability, suitability, and control of water.

Although there are many more demands on water resources, this is true today, just as it has been for millennia. The main difference, however, is that we have come to recognize the interconnectedness among elements in an ecosystem, including humans – genuinely embracing all that is connoted in the term.[3]

Many refer to watersheds as complex socioecological systems (e.g., Barham 2001; Lowe et al. 2003). Quoting Bobbie Lowe and colleagues (2003, 103), Schlager and Blomquist (2008, 3) portray these socioecological systems as complex adaptive systems that "are composed of a large number of active elements whose rich patterns of interactions produce emergent properties which are not as easy to predict by analyzing the separate system components." Due to this complex nature, there is a great deal of uncertainty characterizing these systems (Dietz, Ostrom, and Stern 2003; Imperial 2005; Schlager and Blomquist 2008; Walker, Daniels, and Emborg 2008), which restricts our empirical tools, forcing us to analyze individual components as a larger part of the system or, even better, to view the individual components as interconnected parts of a networked system. Numerous actors are focused on these components, some attempting to act collectively to benefit the entire system.

What do we know about the landscape and the need for collaboration? One part is the recognition that collaborations are complex social and ecological systems,

[3] It must be noted that while this understanding of ecosystem interconnectedness is relatively recent to Western cultures, many indigenous and Eastern cultures have long held this belief.

which "do not need to be simplified for one or two values, their complexity is to be recognized, protected, and respected" (Schlager and Blomquist 2008, 151). The social sciences have long viewed political complexity as a "barrier to sustainable use of natural resources and something to be corrected" (Schlager and Blomquist 2008, 151). These policies seek to simplify, integrate, and centralize so that the externalities, spillover effects, commons dilemmas, and public goods can be traded off and balanced against each other. However, "[a]ll of this complexity cannot be quieted by relying on science or nature to define boundaries or by forcing diversity into the well-designed box of hierarchy" (Schlager and Blomquist 2008, 152). A recommended solution is embracing these complex and adaptive issues with a system similarly composed: federalism (e.g., Feiock and Scholz 2010). "Federalism encourages innovation and experimentation because of the dynamic tension between self-rule and shared-rule" (Schlager and Blomquist 2008, 153).[4]

Another part of the answer lies within the very nature of American environmental federalism and the vexing nature of watersheds (Morris et al. 2014). Historical and more current political trends, for example, favor less antagonistic relationships between stakeholders. Decentralized institutions – ones with official jurisdiction over resources across several states through interstate compacts, for instance – have been guided by congressional legislation and are often dependent on federal financing, and they have had mixed results regarding environmental improvement outcomes. Other cooperative approaches exist, such as the National Estuary Program (NEP), which partners federal, state, and local agencies and often provides a model for watershed collaborations. Watersheds vary significantly in their size, location, priorities, and purposes, meaning that decisions can have cascading and multiplicative impacts that go well beyond traditional jurisdictional lines. As such, managing these impacts and/or restoring or protecting them requires collaboration that involves the public, nonprofit, and private sectors. Thus, intergovernmental relations are pivotal to effective management, as policy implementation necessitates coordination as well as collaboration among impacted stakeholders.

The scope and scale of impaired waters that may necessitate such collaborative approaches is significant:

- The EPA and the states have identified and listed approximately 70,000 waterbodies across the nation that do not meet water quality standards (GAO 2022).

[4] This idea is made explicit in Vincent Ostrom's (1974) University of Alabama lecture series, *The Intellectual Crisis in American Public Administration*.

- While precise estimates are difficult due to data limitations, in 2014, the GAO estimated that 50,000 TMDLs were developed and approved by the EPA.
- In a sample of TMDLs analyzed by GAO, 191 TMDLs reported that pollutants had been reduced in many waters, "but few impaired water bodies had fully attained water quality standards" (GAO 2013, 35).

In their study of TMDL implementation in watersheds in Ohio and West Virginia, Hoornbeek and colleagues (2012, 422) found "that the presence of watershed groups significantly increases the probability of TMDL implementation progress." After examining 63 TMDL reports and talking with knowledgeable state officials about implementation progress and perceived pollutant loading reductions, they conclude that collaborative efforts to implement TMDL agreements are commonplace (Hoornbeek 2012, 425–26). Nonetheless, there were several TMDL-listed watersheds in which they could find no implementation progress.

2.3 Conclusion

The governance of US water quality has evolved significantly. Its earliest days were characterized by a fragmented, wholly incomplete, and locally based system – and water was linked mostly to transportation. As of 2025, a more structured, formal, albeit still complex and fragmented intergovernmental system works to deliver the promise of the Clean Water Act. In fact, this system has enjoyed much success and has largely regulated point-source pollution effectively (Morris 2022). Yet the durable and ongoing challenges of nonpoint source water pollution has demonstrated the limitations of command-and-control regulatory approaches. Additionally, the principles embedded in Sections 303 and 319 of the CWA have encouraged states to adopt governance approaches that moved away from traditional regulation and toward more collaborative and participatory strategies. These factors, among others, set the stage for watershed collaboration – a governance strategy designed to address diffuse pollution sources, resource constraints, and a complicated socio-ecological environment.

Over time, stakeholders have increasingly recognized this promise and have turned to collaborative governance to address impaired waters. Existing empirical evidence remains mixed, as there are thousands of still impaired waters across the United States and still more TMDL programs that have had, at best, limited and uneven success. As a result, collaborative watershed governance offers promise but with unknown efficacy.

References

Association of Interstate Water Pollution Control Administrators. 2004. *Clean Water Act Thirty-Year Retrospective: History and Documents Related to the Federal Statute*. Tech. rept. Association of Interstate Water Pollution Control Administrators, Washington, D.C.: Association of Interstate Water Pollution Control Administrators.

Barham, Elizabeth. 2001. Ecological Boundaries as Community Boundaries. *Society and Natural Resources*, **14**, 181–191.

Crotty, Patricia M. 1987. The New Federalism Game: Primacy Implementation of Environmental Policy. *Publius: The Journal of Federalism*, **17**, 53–67.

Davies, J. Clarence, and Davies, Barbara S. 1975. *The Politics of Pollution*. 2nd ed. Indianapolis, IN: Bobbs-Merrill.

Dietz, Thomas, Ostrom, Elinor, and Stern, Paul C. 2003. The Struggle to Govern the Commons. *Science*, **302**(December), 1907–1912.

Eames, Diane D. 1970. The Refuse Act of 1899: Its Scope and Role in Control of Water Pollution. *California Law Review*, **58**, 1444–1473.

Environmental Law Institute. *Compendium of Water Quality Restoration Approaches*. n.d.

Environmental Protection Agency. 1996. *Watershed Approach Framework*. Tech. rept. EPA 840-S-96-001. Environmental Protection Agency.

Environmental Protection Agency. 2022a (September). *2022–2032 Vision for the Clean Water Act Section 303(d) Program*. Tech. rept. Environmental Protection Agency.

Environmental Protection Agency. 2022b (September). *Clean Water Act Section 303(d) Vision*. Tech. rept. Environmental Protection Agency.

Environmental Protection Agency. 2024. Water Quality Standards Regulatory Revisions to Protect Tribal Reserved Rights. *Federal Register*, **89**(86), 35717–35748.

Ernst, Howard R. 2003. *Chesapeake Bay Blues: Science, Politics, and the Struggle to Save the Bay*. Lanham, MD: Rowman and Littlefield.

Ernst, Howard R. 2010. *Fight for the Bay: Why a Dark Green Environmental Awakening Is Needed to Save the Chesapeake Bay*. Lanham, MD: Rowman and Littlefield.

Feiock, Richard C., and Scholz, John T. (eds.). 2010. *Self-Organizing Federalism: Collaborative Mechanisms to Mitigate Institutional Collective Action Dilemmas*. New York: Cambridge University Press.

Florida Department of Environmental Protection. *Surface Water Quality Standards: Classes, Uses & Criteria*. n.d.

Freeman, A. Myrick III. 1990. Water Pollution Policy. In *Public Policies for Environmental Protection*, edited by Paul Portney, 97–149. Washington, D.C.: Resources for the Future.

Government Accountability Office. 2013 (December 5). *Clean Water Act: Changes Needed if Key EPA Program Is to Help Fulfill the Nation's Water Quality Goals*. Tech. rept. GAO-14-80. Government Accountability Office.

Government Accountability Office. 2022 (October 17). *50 Years After the Clean Water Act – Gauging Progress*. GAO WatchBlog.

Grigg, Neil S. 1999. Integrated Water Resource Management: Who Should Lead, Who Should Pay? *Journal of the American Water Resources Association*, **35**(3), 527–534.

Hoornbeek, John, Hansen, Evan, Ringquist, Evan, and Carlson, Robert. 2008. *Implementing Total Maximum Daily Loads: Understanding and Fostering Successful Results*. Tech. rept. Report details incomplete in source.

Hoornbeek, John A. 2012. *Water Pollution Policies and the American States: Runaway Bureaucracies or Congressional Control?* State University of New York Press.

Houck, Oliver A. 2002a. *The Clean Water Act TMDL Program: Law Policy, and Implementation*. Washington, D.C.: Environmental Law Institute.

Houck, Oliver A. 2002b. The Clean Water Act TMDL Program V: Aftershock and Prelude. *Environmental Law Reporter News & Analysis*, **32**, 10385.

Imperial, Mark. 2005. Using Collaboration as a Governance Strategy: Six Lessons from Watershed Management Programs. *Administration and Society*, **37**(3), 281–320.

Koontz, Thomas M., and Thomas, Craig. 2006. What Do We Know and Need to Know about the Environmental Outcomes of Collaborative Management? *Public Administration Review*, **66**, 111–121.

Lowe, Bobbie, Ostrom, Elinor, Simon, Carl, and Wilson, James. 2003. Redundancy and Diversity: Do They Influence Optimal Management? In *Navigating Social-Ecological Systems: Building Resilience for Complexity and Change*, edited by Fikret Berkes, Johan Colding, and Carl Folke, 83–114. New York: Cambridge University Press.

Lubell, Mark, Schneider, Mark, Scholz, John T., and Mete, Mihriye. 2002. Watershed Partnerships and the Emergence of Collective Action Institutions. *American Journal of Political Science*, **46**(1), 148–163.

Mayer, Martin K., Morris, John C., McNamara, Madeleine W., and Zhang, Xiaodan. 2024. Explaining State Efforts to Create Total Maximum Daily Load (TMDL) Agreements. *Social Science Quarterly*, **105**(5), 1776–1790.

McNamara, Madeleine W. 2016. Unraveling the Characteristics of Mandated Collaboration. In *Advancing Collaboration Theory: Models, Typologies, and Evidence*, edited by John C. Morris and Katrina Miller-Stevens, 65–86. New York: Routledge.

Milazzo, Paul Charles. 2006. *Unlikely Environmentalists: Congress and Clean Water, 1945–1972*. Lawrence, KS: University of Kansas Press.

Mills, Mollie, and Oberthur, Aleena. 2024 (September 5). *Water System Upgrades Could Require More than $1 Trillion Over Next 20 Years*. The Pew Charitable Trusts.

Morris, John C. 2022. *Clean Water Policy and State Choice: Promise and Performance of the Water Quality Act*. New York: Cambridge University Press.

Morris, John C., Gibson, William A., Leavitt, William M., and Jones, Shana C. 2013. *The Case for Grassroots Collaboration: Social Capital and Ecosystem Restoration at the Local Level*. Lanham, MD: Lexington Books.

Morris, John C., Gibson, William A., Leavitt, William M., and Jones, Shana C. 2014. Collaborative Federalism and the Emerging Role of Local Nonprofits in Water Quality Implementation. *Publius: The Journal of Federalism*, **44**(3), 499–518.

Poe, Gregory L. 1995. *The Evolution of Federal Water Pollution Control Policies*. Tech. rept. Department of Agricultural, Resource, and Managerial Economics, Ithaca, NY.

Sabatier, Paul A., Leach, William D., Lubell, Mark, and Pelkey, Niel W. 2005. Theoretical Frameworks Explaining Partnership Success. In *Swimming Upstream: Collaborative Approaches to Watershed Management*, edited by Paul A. Sabatier, Will Focht, Mark Lubell, Zev Trachtenberg, Arnold Vedlitz, and Marty Matlock, 173–199. Cambridge, MA: The MIT Press.

Schlager, Edella, and Blomquist, William. 2008. *Embracing Watershed Politics*. Boulder, CO: University Press of Colorado.

Stern, Duke E., and Mazze, Edward M. 1974. Federal Water Pollution Control Act Amendments of 1972. *American Business Law Journal*, **12**(1), 81–86.

USGS. 2012. *Water Resources of the United States: Hydrologic Unit Maps*.

Walker, Gregg B., Daniels, Steven E., and Emborg, Jens. 2008. Tackling the Tangle of Environmental Conflict: Complexity, Controversy, and Collaborative Learning. *Emergence: Complexity and Organization*, **10**(4), 17–27.

Wayland, Robert. 2012. The Tortuous Road to 'No Net Loss' of Wetlands. In *True Green: Executive Effectiveness in the U.S. Environmental Protection Agency*, edited by Gerald Andrews Emison, and John C. Morris, 85–109. Lanham, MD: Lexington Books.

3

The Collaborative Approach

Watersheds are complex adaptive systems. This complexity necessitates institutional arrangements that are agile and adaptive enough to effectively manage such a system. In other words, just as watersheds are complex and adaptive, the relevant institutional arrangements must also exhibit complexity and adaptability. Yet many institutional arrangements struggle to effectively manage such systems. In fact, many conflicts within watersheds result from collective action problems, which themselves stem from institutional and ecological complexity (Scholz and Stiftel 2005). As a result, some stakeholders have championed more adaptive governing institutional arrangements, whereas others have sought more collaborative approaches as a way for self-governing resource users to deal with complexity and uncertainty (Ostrom 1990; 1999; Ostrom, Gardner, and Walker 1994).

Several common design elements are apparent when referring to collaborative governance. For Barry Rabe (2009, 49), these elements include coordination and integration among participants, along with the ability to develop suitable metrics to monitor performance. Other common characteristics include their "outward-looking" nature (Kettl 2009, 10), goal congruence (Rabe 2009, 51), the ability to manage knowledge and information (McNamara 2012; Agranoff 2007; Hale 2011; Rabe 2009), boundary crossing and boundary spanning organizations (Agranoff 2007; Moore 2009), stakeholder involvement (Morris et al. 2013; Agranoff 2008; Cooper, Bryer, and Meek 2008; Durant, Fiorino, and O'Leary 2004; Hale 2011; Jordan and Wolf 2006; Sabatier et al. 2005), and various decision-making strategies and arrangements (Agranoff 2007; Imperial 2005; Schlager and Blomquist 2008).

Attempts at addressing these collective action challenges vis-à-vis watersheds have acquired numerous labels and theoretical approaches. Regardless of whether they are identified as collaboratives, networks, or polycentric arrangements, these efforts can be described as self-organized organizations due to the numerous actors involved at multiple levels (Feiock and Scholz 2010). Political realities often complicate these approaches, and in the end, "people are not free to design any type of institution or policy they desire: they must settle on a tolerable mix of transaction costs" (Schlager and Blomquist 2008, 98). Moreover, because watersheds exist at multiple scales and have multiple intended uses, watershed management requires participants to pursue simultaneously multiple goals, which may or may not be complementary (Schlager and Blomquist 2008, 127).

In many places collaborations and networks of various types overlap at different scales and are pursuing distinct goals. What we have, then, is a diverse multi-organizational, noncentralized (Elazar 1987), polycentric (E. Ostrom 2005) institutional arrangement within a watershed that

> results from the combined effects of differences among the interests of individuals and organizations within the watershed, changes over time in the understanding of problems and in values for water, the need to pursue multiple resource management goals, distinctions in the efficient scale of operation of organizations performing diverse functions, and the importance of finding distributions of benefits and costs that are perceived as fair by those who bear them. (Schlager and Blomquist 2008, 148)

This is where the actual political and vexatious nature of watershed governance is most evident. However, it also creates opportunities for resource users and stakeholders to come together in pursuit of these diverse goals, attempting to overcome the collective action dilemmas that are inherent in such an endeavor (Olsen 1971; Ostrom 1990).

3.1 Why Pursue Collaboration?

A benefit of collaborative governance is its adaptive capacity (Sabatier et al. 2005). Vincent Ostrom observed that managing water resources in the United States requires institutional arrangements that recognize both the diversity and interdependence of interests within river basins (V. Ostrom 1962b, 65). The challenge arises from the multiple uses and users within a single watershed (V. Ostrom 1962a, 450). Effective mechanisms must adapt to changing natural processes and social circumstances, ensuring all stakeholders are included in the process.

Collaborative governance structures are designed to address the limitations of human and natural systems and to adapt to likely responses and surprises from the natural system, often through best management practices (BMPs) such as adaptive management and continuous learning. While specialized command-and-control agencies have addressed first-order collective action problems, their success can lead to new conflicts as decisions by one authority affect others (Scholz and Stiftel 2005, 1). Second-order collective action problems require changes in governing institutions to strengthen adaptation while respecting agency capabilities and authority (Scholz and Stiftel 2005, 5).

Another benefit of collaboration is its ability to mediate, manage, and resolve conflict. Scholz and Stiftel (2005) argue that adaptive governance is particularly useful in collaborative policy networks, such as those formed around watersheds, as a mechanism to manage conflict effectively. Collaboration addresses conflict in ways that enhance joint gains, reduce transaction costs, and promote sustainable resource systems through various forms of conflict resolution.

> Conflicts that arise under conditions of stress on the natural system frequently arise because of unexpected responses from the natural system and require affected agencies to deal with unfamiliar issues beyond their established expertise. They frequently involve new stakeholders, unfamiliar with the agencies' procedures... Coordinating policies across fragmented arenas could produce considerable benefits for stakeholders jointly affected by the decision of specialized authorities, but this requires combinations of expertise, authority, and representation of users that are not yet an established part of the institutional structure governing water resources (Scholz and Stiftel 2005, 3).

Scholz and Stiftel (2005) use examples from water conflicts in Florida to illustrate the issues of the fragmentation of political authority, the role of conflict, and the use of pragmatic approaches to collective problem solving. Some institutional innovations, like the National Pollution Discharge Elimination Systems (NPDES), have enjoyed success at reducing point-source pollution and have resolved many collective action problems and conflicts. Yet others have been unable to mediate conflicts and, in some cases, have heightened the level of conflict (Scholz and Stiftel 2005). When these second-order conflicts become intractable, the potential of second-generation environmental institutions that self-organize, learn, and grow is most pronounced (Durant, Fiorino, and O'Leary 2004; Feiock and Scholz 2010; Lubell et al. 2002; Sabatier et al. 2005) as a mechanism to manage conflict effectively.

The extant collaboration literature addresses a number of concepts and other theoretical elements that serve to frame our understanding of watershed collaboration. The following sections of this chapter introduce these elements; the latter part of this chapter presents a framework we use to test our propositions about the outcomes of collaborative watershed management.

3.2 The Context of Collaboration and Watersheds

Garett Hardin's (1968) classic assessment of *The Tragedy of the Commons* provides general context to managing environmental and natural resources. Some may view the conflict and problems inherent in water resource management as a tragedy when competing users increase resource consumption in the face of limited resources. Hardin (1968, 1244) wrote, "Freedom in a commons brings ruin to all." Responding to this warning several decades later, however, Dietz, E. Ostrom, and Stern (2003, 1907) argue that Hardin failed to account for the ability of self-governing institutions to protect against threats of resource scarcity. That is the point that proponents of self-organized governance (and self-organized federalism) attempt to make: It is a struggle, but not necessarily a tragedy. Dietz, Ostrom, and Stern (2003, 1907–8) continue,

> Devising ways to sustain the earth's ability to support diverse life, including a reasonable quality of life for humans, involves making tough decisions under uncertainty, complexity, and substantial biophysical constraints as well as conflicting human values and interests. Devising effective governance systems is akin to a coevolutionary race. A set of rules crafted to fit one set of socioecological conditions can erode as social, economic, and technological developments increase the potential for human damage to ecosystems and even the biosphere itself.

Numerous studies have demonstrated that commons users are up to the challenge and can devise institutions to solve their dilemmas (e.g., Blomquist 1992; E. Ostrom 1990; E. Ostrom, Gardner, and Walker 1994; Schlager 2002). Of course, not all watersheds are considered common pool resources (CPR), and CPR theory does not directly apply to these alternate contexts. However, Feiock and Scholz (2010) demonstrated that many of the challenges and solutions appear to be quite similar.[1] Of course, there may always be tragic choices among competing users, but recognizing this tragedy is the first step to effective policy that can benefit both natural and human systems (Adams 2003). This recognition may be an essential early step to developing agile institutional arrangements that can successfully, effectively, and sustainably manage watersheds. The first step, however, is to recognize applicable stakeholders, the community's socioeconomic conditions as well as the ecological attributes that define the watershed.

3.2.1 Stakeholders and Socioeconomic and Community Conditions

Watershed stakeholders are heterogeneous in many ways. They range from, for example, power companies that operate hydroelectric power-generating dams, to homeowners who live in a watershed, to resource users who provide water to communities, to fisheries, industries, and governments at various levels, to conservation groups, to recreational users, to agricultural users, to sporting groups, to academic researchers, and specialist practitioners. Sabatier and colleagues (2005, 20) remind us that stakeholders are also

> policymakers, agency implementers, experts both within and outside government who participate in policymaking and policy implementation, private sector businesses that are

[1] Although the institutional landscapes were far more sparse in E. Ostrom's (1990) award-winning study of common pool resource dilemmas, many of the same conclusions were reached regarding the effect of self-organizing governance systems on autonomy, monitoring, enforcement, compliance, and legitimacy.

economically or otherwise affected by policies, members of the general public who are economically or otherwise affected by policies, and environmental interest groups that purport to represent nonhuman values, among other groups.

In some instances, traditional battle lines are drawn between stakeholders, whereas in others, they break down as former adversaries become partners (Morris et al. 2013; Kash 2008). A key variable in distinguishing conflictual partnerships from more harmonious ones is the development of trust. In this context, trust can foster the requisite commitment and dedication necessary for successful collective action among watershed stakeholders (Lubell 2007; Scholz, Berardo, and Kile 2008). This dynamic is especially important (Lubell 2007) if commitment among stakeholders is of primary concern (Scholz, Berardo, and Kile 2008).

Other scholarship has focused on the composition of the collaboration or network. Ryan Bidwell and Clare Ryan (2006), for example, show that diverse stakeholder participation in watershed partnerships is necessary to achieve changes in water resource management. However, more homogeneous or smaller groups may be more likely to conduct watershed improvement activities (Morris et al. 2013; 2014).[2]

Beyond the relevant stakeholders, other community attributes affect how the resource is used and managed. Watershed partnerships are more likely to form in more densely populated areas, and more homeowners within a watershed are likely to increase the available social capital required to collaboratively manage the resource (Hoornbeek et al. 2012; E. Ostrom 1998; E. Ostrom, Gardner, and Walker 1994). This point was echoed by Morris et al. (2013), who found that social capital is not only a product of collaboration, but that some level of social capital is a necessary precondition. Stakeholder participation is also more likely among those with greater socioeconomic resources (Koehler and Koontz 2008). Finally, higher levels of political efficacy facilitate citizen participation and engagement in public policy debates and in collaborative decision-making processes (Hoornbeek et al. 2008).

3.2.2 Ecological Conditions

Understanding the ecological components and conditions of the natural resource is an essential step in understanding how the watershed can be affected by social systems. In the first place, this can determine the local economy and intended uses: Components and conditions may determine if the primary economic engine of a watershed's community is tourism, mining, aquaculture, shale gas extraction, etc.

[2] Mancur Olsen (1971) described why this may be the case in his well-known monograph, *The Logic of Collective Action*.

Among these ecological conditions are differences in scale, water flow characteristics, inherent uncertainties and complexities of natural resource systems, and the heterogeneous geographic, geologic, topographic, hydrologic, and hydrographic features of watersheds.

As an ecological condition, stream flow is considered a "master variable" when understanding how riverine processes work. It is "strongly correlated with many critical physiochemical characteristics of rivers, such as water temperature, channel geomorphology, and habitat diversity" (Poff et al. 1997, 769). It also "limits the distribution and abundance of riverine species … and regulates the ecological integrity of flowing water systems" (Poff et al. 1997, 769). Humans have altered natural flowing waters throughout history with dams, channels, levees, groundwater pumping, and the effects of urbanization. All of these have had a significant impact on natural watersheds and have changed the ways these watersheds work. Across the United States, especially in heavily altered watersheds, there are calls to return rivers to a natural flow regime that mimics the river's original state before alternation. Removing dams is one way to do this. Still, many more ways require a community of actors to come together in a collaborative fashion to meet the needs of human stakeholders and the watershed's biotic integrity (Irwin and Freeman 2002; Jacobson and Galat 2008; Poff et al. 1997).

No matter how a watershed is managed, the river system therein is governed by some flow regime; whether it be altered, natural, or natural-mimicking, some group of individuals and/or institutions determine how the water flows. These flow processes are important to all water resource users. To fisheries stakeholders, for example, flow determines the ability of eggs to attach to riverine substrate. For agriculture stakeholders, as another example, flow can determine the richness of alluvial soils. In short, flow, in general, is important to all stakeholders. When a river system is not in its natural state, and when the circumstances are suitable for long-term collaborative commitment by interested stakeholders, flow can be governed by adaptive management. This approach has been recommended in many circumstances by the EPA, the Department of Interior,[3] and other various land and water management agencies.

3.2.3 Uncertainty

Complexity and uncertainty are endemic to watershed institutions and stakeholders. These realities come not only from the biophysical nature of the system but also from complex and uncertain social aspects among those involved in its

[3] In fact, the Department of Interior has developed numerous resources and technical guides to assist stakeholders in the implementation of adaptive management processes. This can be found at www.usgs.gov/publications/adaptive-management-us-department-interior-technical-guide.

management (V. Ostrom 1962b; Walker et al. 2008). Uncertainty is related to the differing rates of change of various physical watershed aspects, scale differences, near decomposability, and numerous disturbance processes (Blomquist 2009). Complexity is compounded by these uncertainties, leading to inaccurate predictions about the system and disagreement about which system elements best indicate system conditions (Blomquist 2009). To further explain the difference between complexity and uncertainty, Blomquist (2009, 112) quotes Roe (2001, 111): "Issues are uncertain when causal processes are unclear or not easily understood. Issues are complex when they are more varied and interrelated than before." These issues of complexity and uncertainty are often dealt with through scientific learning, a guiding principle of adaptive management. A greater scientific understanding of the nature of natural resources and their causal processes enhances the ability of stakeholders to acknowledge and understand sources of complexity and uncertainty (Blomquist 2009; Roe 2001; Walker, Daniels, and Emborg 2008).

3.2.4 Decision-Making Processes

Regardless of settings, complexity, ecological conditions, and context, or perhaps because these aspects vary across landscapes, collaboration has been defined in many ways, often focusing on the process or participants and how they are situated within an institutional structure. Chris Ansell and Alison Gash (2007) attempt to distinguish between the "species" of collaboration – watershed councils, community policing, community health partnerships, ecosystem comanagement, and others – from the "genus" of collaborative governance. Identifying some essential criteria, Ansell and Gash (2007, 544) define collaboration as a

> governing arrangement where one or more public agencies directly engage nonstate stakeholders in a collective decision-making process that is formal, consensus-oriented, and deliberative and that aims to make or implement public policy or manage public programs or assets.

Other definitions abound, including some that capture formal and informal collaborative activities and relationships. Donna Wood and Barbara Gray (1991) profoundly influenced the development of collaboration scholarship by calling for a standard, more complete definition when theorizing about its processes. Following their lead, Thomson, Perry, and Miller (2007, 25) define collaboration as a

> process in which autonomous or semi-autonomous actors interact through formal and informal negotiation, jointly creating rules and structures governing their relationships and ways to act or decide on the issues that brought them together; it is a process of shared norms and mutually beneficial interactions.

This definition highlights five critical dimensions of collaboration: governance, administration, organizational autonomy, mutuality, and norms (Thomson, Perry, and Miller 2007), two of which are examined here.

The governance dimension does not necessarily imply everyone in the collaboration has to agree on the best possible solution; instead, the goal is for participants to support decisions reached by the group (Thomson and Perry 2006, 24). Thomson and Perry (2006, 24) suggest that successful collaboration is built on the assumption that participants are not only willing to monitor the group (including themselves) but are also willing to impose sanctions on participants who are not compliant with the will of the group. As this cycle repeats, collective responsibility is fostered, moving participants from problem-blaming to problem-solving (Thomson and Perry 2006, 25). Sabatier et al. (2005) suggest that a successful collaboration can adapt to changing environmental, political, and social circumstances. In many instances, partnerships and collaboratives with adaptive capacity are necessary to address the changing and interconnected nature of watersheds and implement programs and policies designed to improve or sustain environmental conditions.

Studies have also addressed the administrative dimensions of collaborative decision-making. Thomson and Perry (2006, 25) observe that good communication, achievable, concrete goals, clear roles and responsibilities, and boundary setting fostered better collaborative environments and decisions. Short-term successes hold the collaboration together; otherwise, it appears it can fall victim to institutional inertia. More importantly, collaborative members should represent those living and working within the watershed; This enhances partnership legitimacy – giving credence to the administrative goals – and develops a civic community at the heart of self-governing.

In short, as Menzel (1987) acknowledges, the complexity of collaboration makes it unlikely that a single theory of collaboration can explain all aspects of the interaction, Imperial (2005) notes that there are many ways to structure collaborative activities. Imperial (2005) adapted the Institutional Analysis and Development (IAD) framework to create his Levels of Collaborative Action Framework (LCAF) in his analysis of six watershed management programs. While the IAD framework analyzes rules at the three levels discussed previously, the LCAF acknowledges that collaborative activities at each level are interrelated (Imperial 2005, 288). With this view, there are three levels of joint action: operational, policy making, and institutional; these respectively correspond to the levels of the IAD framework: operational, collective choice, and constitutional. It is worth noting that McNamara (2012) took issue with the simplicity of Imperial's approach, noting that Imperial's (2005) framework was so general that it lacked the ability to distinguish between different forms of interaction. However, this serves to reinforce Menzel's (1987) basic point that developing useful collaboration theory is difficult.

3.2.5 Information and Stakeholders

Agranoff (2012, 23) suggests that collaborative agreements and networks are often supported and kept together via information and goals. The most similar – some say most important, however formal or informal – is the engagement of stakeholders and the value of strategic information and goal setting (McNamara 2016a; Gazley 2008, 40). Hale (2011, 9) notes that intergovernmental systems in the United States, as commonly conceived, are, in effect, networks of interdependent actors that include both government and nongovernmental participants. Heilman et al. (2024) noted the same feature in water policy networks; they also report that water policy networks are also arranged into two major subnetworks (water quality and water quantity) and include actors from across the intergovernmental spectrum as well as participants from different sectors. These networks are comingled at several levels and work at different stages of the policy process. These are the relationships we see in watershed management. When these stakeholder relationships are well organized, Agranoff (2007, 2) describes them as public management networks. On the other hand, McNamara (2016b) suggests that collaboration should be thought of as something other than public management, even though participants in collaboration can, and often do, include governmental stakeholders. Moore and Koontz (2003) define three types of collaboration: agency-based, citizen-based, and mixed. The existence of a citizen-based form suggests that collaboration can exist independently of government. An important distinction here is the role of government and if the collaboration can exist independently of governmental actors.

As discussed so far, we know much about collaborative and network formation, characteristics, types, potentialities for success, and other information. Still, we know very little about the actual impacts to stakeholders operating in a water policy arena. Lubell (2005, 223) examined estuaries around the country that were part of the National Estuary Program (NEP) and those that were not to determine if watershed partnerships like the NEP impacted stakeholder beliefs to increase cooperation, and thus policy effectiveness. He found that changes in collective action beliefs among stakeholders in NEP estuaries suggested that the NEP decreased transaction costs and increased benefits in the watershed (Lubell 2005, 223). Numerous case studies of watershed collaboratives have reached similar conclusions, leading to consensus among scholars that the characteristics of the collaboration itself are likely to increase the perceived effectiveness of that collaboration. Common metrics linked to this relationship include the number of organizations or representatives involved in the collaborative or network, the size of the collaborative's membership (i.e., number of staff and volunteers in the member organizations), and the amount of resources the collaborative can secure (Hardy and Koontz 2009; Heikkila 2004; Sabatier et al. 2005). Perceived effectiveness, perceived outcomes, and other perceptions are all the evidence we have to say these types of partnerships lead to positive environmental outcomes. Of course, these perceptions can be the result of confirmation bias (see Kahneman 2011, 80–81).

3.2.6 Operational Activities and Outcomes

There is a general understanding of why watershed partnerships may form, what they do, who is involved, and how they survive. Evidence shows that they produce numerous perceived social benefits and generate various administrative outputs (see Morris et al. 2013; Sabatier et al. 2005). Little is known, however, about the actual outcomes they produce from collaborative policy implementation. Moreover, there is no consensus about whether they are more effective than the traditional command-and-control institutions they supplement or replace. However, the literature is replete with studies that note the limitations of identifying and assessing watershed outcomes (Koontz and Thomas 2006).

If outcomes measure the success of watershed collaboratives, there are several notable challenges, according to Koontz and Thomas (2006, 113–14), including:

1. Gathering data that accurately reflects environmental outcomes.
2. Accounting for the often lengthy timeframes required for collaborative outputs to result in observable environmental changes.
3. Developing research protocols capable of disentangling the effects of multiple, interacting variables influencing environmental change.
4. Creating large-*n* studies and datasets that enable outcome measurement across numerous collaborative efforts.

The question at hand, then, is how to make valid inferences about the effects of collaboration. Though numerous books and journal articles present evidence of collaborative environmental outcomes from case studies, generalizability remains limited. One productive approach to advancing this line of inquiry draws from institutional and collective action theory, emphasizing three core domains: (1) The attributes of the resource, (2) the characteristics of the institutions governing it, and (3) the social and political context of the community and stakeholders involved (Lubell et al. 2002; E. Ostrom 1990, 2005; Ostrom, Gardner, and Walker 1994).

3.2.7 Defining Success

One of the enduring issues in watershed collaboration research is the question of how to define "success" in collaborative efforts (McNamara et al. 2020; Morris and Miller-Stevens 2016). Arguably, the long-term goal of many watershed collaboration efforts is grounded in restoration: to return the water quality in a watershed to a prior (cleaner) state. Other efforts might be targeted toward watershed protection: to prevent a degradation of water quality in a watershed. Still other efforts might attempt to increase public recreational access to water resources to prevent industrial or residential development in a watershed, or other forms of action.

One way to operationalize effectiveness is by whether collaborators can successfully develop projects and, in the long run, show improvements to the watershed (Sabatier et al. 2005, 179). This emphasizes operational level activities and outputs. This may be why Imperial (2005, 289) recognizes that the focus of collaborative activities is rightly on the actual activities of the group. Actions at this level are designed to (1) directly improve environmental conditions, (2) educate the public and decision-makers, and (3) monitor conditions and enforce regulations.

The question of "success" in collaboration has also been addressed by Morris et al. (2013) and McNamara et al. (2019). For Morris et al., success in collaboration can be defined by an ongoing process. In other words, if there is activity continued, collaboration can be defined as successful. They note that an ongoing process could increase social capital, which in turn is an indicator of collaboration success. Likewise, collaborative processes can also generate short-term outputs (e.g., plans, projects) and long-term outcomes (completed activities that lead to measurable changes in the watershed).

On the other hand, McNamara et al. (2019) attempt to address "failure" in collaboration. They note that while the available literature is rife with examples of successful collaboration, little is known about the reasons for collaboration failure. They conclude that collaborative efforts tend to fail, not because of a lack of outputs or outcomes, but rather because of issues related to organizational dynamics, interorganizational processes, or poor management. Still, it makes intuitive sense to suggest that measurable outputs and outcomes may contribute to the longevity of a collaborative effort.

One might also posit that collaboration success may be defined through longer-term goals. Here, the goals of the effort can be measured in scientific terms through careful testing and monitoring of pollutants in the waterway. Indeed, such testing is an integral part of the Clean Water Act and the Total Maximum Daily Loads process, and both state agencies and private groups (see Morris et al. 2013) engage in water quality testing. In this form, the goal is to detect a long-term decline in pollutants, an increase in overall water quality, and the ability to meet the watershed's intended uses. In the case of Lynnhaven River NOW, a collaborative effort in Virginia Beach, Virginia, goals included a desire to eat oysters from a waterway that was too polluted (historically) to allow human consumption of shellfish. Here, the goal involved a series of actions by the collaborative group to prevent deterioration of water quality. In this case, the methodology is one of regular and ongoing water quality testing, and the measure of success is to see a steady quality in the ambient waters.

In the case of Lynnhaven River NOW, "success" in the collaborative effort is only detectable over time, and only through a rigorous testing regime that tracks water quality over time. In other words, the real effects of collaboration can only be measured in the long term. Some efforts can see measurable gains in short periods of time, but these instances are both rare and limited in geographic scope. Another example is the Elizabeth River Project. Here, stakeholders undertook a project to restore an area of the Elizabeth River in Norfolk, Virginia known as Money Point, which had been the site of significant creosote spill decades before. The project removed thousands of cubic yards of contaminated river bottom and covered the damaged area with clean sand and silt. Within a period of six months scientists noted a dramatic decrease in cancerous growths of a small fish, the Mummichog, that were found at that location in the river.[4] Water quality tests also confirmed a significant decrease in heavy metals and other contaminants at Money Point (Morris et al. 2013). However, the effects were limited to a short segment of the river adjacent to Money Point; the water quality in the rest of the river remained unchanged. Thus, a significant limitation of defining "water quality improvement" as the sole measure of collaborative success is that the effects are likely to be geographically limited, and probably undetectable for long periods of time following the action.

In their book on citizen-based collaboration, Morris, et al. (2013) offer an alternative definition to measure collaborative success. They argue that a successful collaborative process will produce both short-term outputs and long-term outcomes (see Figure 3.1). Outputs can be thought of as agreements, watershed plans, or projects (e.g., remediation of a segment of a waterway; an effort to restore underwater grasses or oyster beds; best practices agreements; or even the implementation of a public education and awareness program). The outputs are thus the decisions that the collaboration makes, and the implementation of the plans and

[4] The Mummichog is a small fish that lives in brackish waters. It differs from many species of fish in that it spends its life very close to where it was hatched. In this case, the Mummichogs in the vicinity of Money Point had shown very high incidences of cancers and other deformities, while Mummichogs 400 yards up or down the river had much lower incidences. After the restoration process, scientists began to notice a nearly immediate drop in the number of affected Mummichogs at Money Point (ERP 2012).

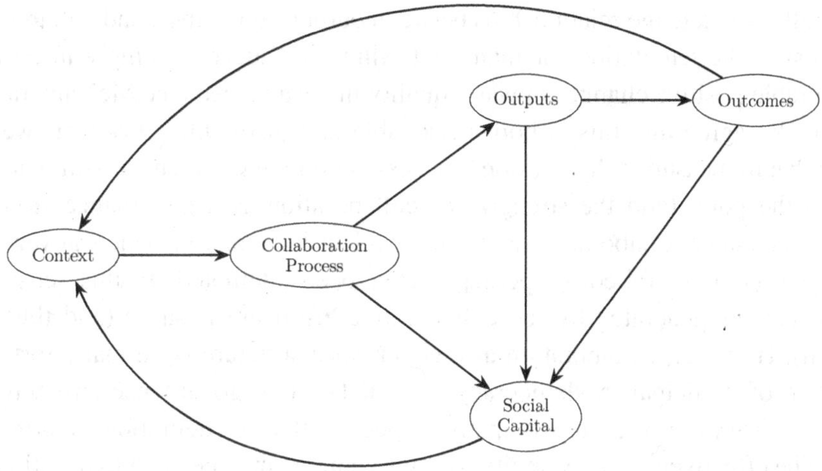

Figure 3.1 Collaboration framework illustrating the dynamic relationship among context, process, outputs, social capital, and outcomes.
Note: Arrows represent causal or reinforcing relationships across components in the collaboration framework. Adapted from Morris et al. (2013).

projects from those decisions. Morris et al. (2013) also argue that an important output of a collaborative process is increased social capital, an important element of collaboration (see Morris et al. 2014; Anderson 2011; Leahy and Anderson 2010; Thomson and Perry 2006). However, social capital serves to enhance future collaboration; it does not directly affect water quality. In short, outcomes are defined as the long-term effects of the outputs (Morris et al. 2013). To this end, we do not address social capital as an output in this study as our focus is directly on collaboration to achieve water quality gains.

An education program to teach children about river stewardship exemplifies the relationship between outputs and outcomes. For example, an educational program is an output that may not immediately improve water quality, but if it successfully changes behaviors, it can lead to measurable long-term improvements (an outcome). Similarly, implementing a BMPs program is an output, while the resulting water quality improvements may only become evident months or years later (an outcome). A social marketing campaign to persuade people to use less lawn fertilizer and prevent runoff into nearby waterways (McNamara et al. 2010) can improve water quality, but the measurable outcomes are not likely to be immediately apparent. Thus, outputs may be thought of as the actions that lead to outcomes. A watershed restoration plan, or even a project in the plan does not itself result in cleaner water, but the successful implementation of a plan or project can, over time, lead to positive water quality changes. Outputs are thus an important precondition of collaboration outcomes. Later in this volume we will address the link between outputs and outcomes in greater detail.

Finally, we note the relationships between outputs, outcomes, and collaboration "success." The temptation for many is to think of "success" simply in terms of measurable positive changes in water quality in a watershed (see McNamara et al. 2020). We agree that this is both reasonable and justifiable. However, we also argue that to measure collaboration "success" purely in terms successful outcomes misses the point (and the strength) of collaboration as a governance structure. First, watershed collaboration is often found in locations in which government interventions have failed, or perhaps never been attempted. In this sense, any improvements generated by the collaborative effort are positive (and thus successful). However, collaboration as a governance structure is, in many forms, an example of participatory democracy. A collaborative governance structure that makes decisions through a participative process is thus, by definition, creating outputs. The effectiveness (or veracity) of those outputs may be open to question, but we suggest that the literature on watershed collaboration would consider the ability to make collective decisions an indicator of success.

Second, because collaborative governance is a form of participative democracy (albeit localized), collaboration can strengthen community structures and interpersonal ties. Often referred to as social capital (see Putnam 2000; Floress et al. 2011), these ties can serve to enhance future participatory activities. Coupled with shared goals (Gray 1989), a successful decision can create buy-in for implementation of the decision, which in turn increases the chances for more impactful outcomes. Third, the old saying that "everyone loves a winner" applies not only to people who are part of the collaborative process, but can also influence those who were not initially a part of the effort. Case studies in the literature are rife with evidence to suggest that the ability of a collaboration to produce decisions (outputs) incentivizes others to participate. In many ways, success breeds success.

This may be particularly important at the local level. Morris et al. (2013) noted an important corollary to social capital: the BIMBY effect. BIMBY (*Because It's My Back Yard*) is a useful concept to understand the motivation of individuals to participate in collaborative watershed governance. It represents a personal link to place, coupled with a notion that there is both a duty and responsibility to take collective action for the collective good. In a re-analysis of the case study data from Morris et al. (2013), Diaz-Kope et al. (2024) found strong evidence not only for BIMBY as motivation for participation, but also as an explanator for successful outputs. More interestingly, the same logic applies to organizational participation in collaboration (Diaz-Kope and Morris 2022; Diaz-Kope and Morris 2019; McNamara and Morris 2021).

We thus suggest that discussions of the results of collaboration are best conceived as being a mix of short-term and long-term indicators. While it may be tempting to attempt to classify collaborations by both indicators, the real world of watershed collaboration is more complicated. First, one must consider the nature of the problem to be addressed. A desire to protect a small body of water in a rural, single political jurisdiction may not require complicated, multiple outputs (or outcomes) to achieve the goal. On the other hand, a watershed with complex water quality issues, and that crosses several political jurisdictions, may present a different level of challenge. In addition, goals can differ markedly from collaboration to collaboration. The goal of watershed protection is not just found in relatively pristine waters; it can also be a goal to prevent a highly polluted waterway from becoming more polluted. Likewise, "restoration" problems can be relatively benign and simple, or remarkably complex and expensive.

Because of these factors, we suggest that any effort to think about collaboration "success" needs to be approached with a clear understanding of the complexities of the problem. There are many examples in the literature of case studies that report successful efforts measured in terms of water quality gains – the outcomes. Campbell, Koontz, and Bonnell (2011), Leach and Sabatier (2005), Lubell (2004), and others (see Koontz and Thomas 2006; Mandarano 2008) focus more on decisions and activities – the outputs. The tremendous variability in collaborative goals, circumstances, watershed problems, etc., makes a large-*n* study such as this challenging. However, we contend that the underlying question – whether collaboration can make a difference in water quality outcomes – remains an important question. Despite limitations in available water quality data and other challenges, there are no large-scale studies in the existing literature that assess the extent to which voluntary collaboration influences water quality. Later in this volume, we will also explore the concept of "success" as reflected in the achievement of collaborative outputs.

3.3 Toward a Model of Collaboration

We now turn our attention to presenting a model of watershed collaboration. This model, informed by relevant collaboration studies, interconnects several key elements: external influences, contextual factors (including ecological characteristics and social capital), information communities, stakeholder community characteristics, legitimacy of the process, policy outputs, and both observed and perceived policy outcomes. The questions and propositions derived from this model are designed to test both the measured and perceived outcomes achieved by those involved in watershed management and governance. In this framework, the watershed serves as the primary unit of analysis.

Figure 3.2 graphically displays our anticipated relationships and related interactions. Many of these relationships are bidirectional, and policy outputs and outcomes are likely to be influenced by several key elements, as described below:

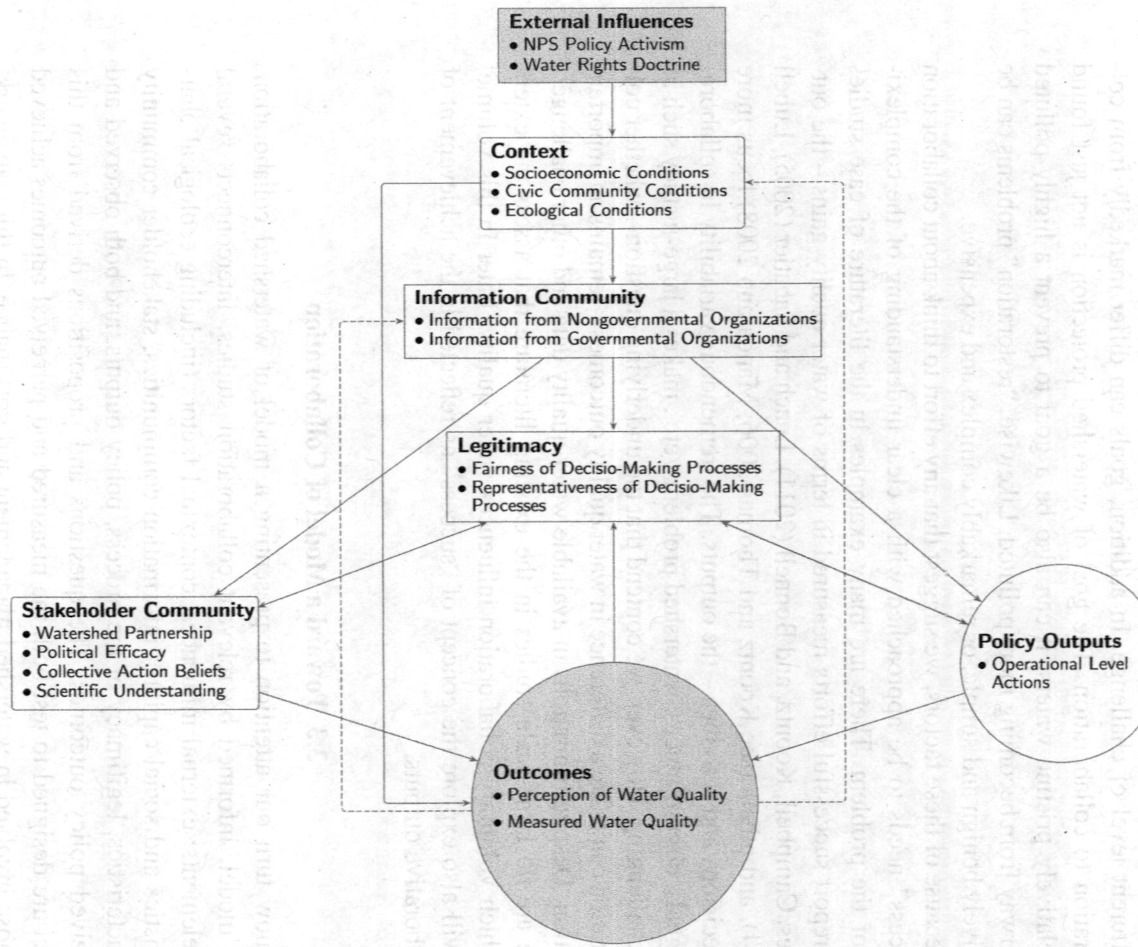

Figure 3.2 Policy framework for watershed governance showing relationships between external influences, community stakeholders, and policy outcomes. Rectangles represent processes, contextual conditions, and legitimacy factors; circles denote outputs and outcomes. Dashed arrows indicate theorized but unmeasured relationships.

- **External Influences:** This represents the overall external environment, for example, political, economic, and/or social variables in which the collaborative efforts are developed and implemented (or not) and shapes the context within which the watershed collaboration exists.
- **Context:** This denotes the relevant civic and ecological conditions within the watershed. These elements shape the nature and operation of the information communities which, in turn, influence the stakeholder groups involved, perceived legitimacy, and the specific outputs of the collaboration.
- **Information Communities:** This describes the mechanisms and platforms in which actors and institutions share information and exchange knowledge, which directly shape legitimacy, the actors involved, and policy outputs.
- **Stakeholder Characteristics:** This is characterized as the attributes of the various stakeholder groups involved. It can include variables approximating stakeholder goals, previous experiences and levels of trust, learning, stakeholder needs, and priorities of the affected groups. Stakeholders here feed into the legitimacy and effectiveness of real and perceived outcomes.
- **Legitimacy:** These are depicted as the trust and acceptance of decisions and actions made by stakeholders. It can also be considered related to transaction costs.
- **Policy Outputs and Outcomes:** These are defined as operational decisions and actions such as BMPs, educative activities, and public outreach taken by stakeholders and are described in more detail later.
- **Outcomes (Observed and Perceived):** This is best described as the tangible and subjective results of the collaboration, including measures of overall water quality and/or creation of trust and other social benefits.

In Figure 3.2, solid arrows represent direct, empirically observed or theoretically established relationships, while dashed arrows indicate feedback loops or theorized connections that are hypothesized but not directly measured in the present analysis. Rectangles are used to denote processes, institutions, or contextual influences that shape collaboration, whereas circles are reserved for outputs and outcomes, emphasizing their role as resulting effects within the system. This visual structure reflects our effort to distinguish between contributing factors and the tangible or perceived results of collaborative watershed governance.

The remaining sections of this chapter explicate our framework and explain how we view collaborative action at the watershed level. The purpose of this framework is to order our concepts in a manner that helps explain the collaborative activity across US watersheds, and to put outcomes from these activities in perspective.

Table 3.1 *Operational level outputs reported by watershed partnerships, grouped by environmental and educative activities.*

Environmental improvement activities	Educative activities
Habitat restoration projects	Public ed. programs for schools
Clean-up activities	Public ed. programs for homeowners
Land acquisition	Public ed. programs for industry
Installing urban best management practices (BMPs)	Public ed. programs for other users
Installing agricultural BMPs	Public social events
Installing forestry BMPs	Special events for the public
Installing water flow BMPs	School curriculum development
Installing sewers	Training programs
Upgrading on-site sewage disposal systems	Special events for elected officials
	Special events for public administrators
	Outreach programs and events

Note: Adapted from Imperial (2005).

3.3.1 The Model and Policy Outputs and Outcomes in Practice

Several policy outputs are listed in Table 3.1. These outputs include BMP installation, habitat restoration projects, educational programs for public officials, school curriculum development, and data collection on environmental activities, among others. If watershed management programs, including stakeholder partnerships, are to improve environmental conditions, it would be through these operational level activities (Imperial 2005; Sabatier et al. 2005). This list provides the operational level activities Imperial (2005) discovered in his investigation of six watershed management programs.

As our model acknowledges, operational activities, such as knowledge sharing, resource sharing, and shared policies, regulations, and social norms, as identified by Imperial (2005), "improv[e] communication among actors, coordinat[e] actions, and integrat[e] policies" as well as "enhance government service delivery" (Imperial 2005, 293). At the institutional level, there are activities that define the parameters of acceptable actions at both the policymaking and implementation levels (Imperial 2005, 298). These activities may also include the institutionalization of shared policies as well as the development of collaborative structures, including either individual members or representatives of other organizations, that act, in effect, as a single organization (Imperial 2005, 299; see also Williams et al. 2016).

3.3.2 Empirical Foci for Watershed Outcomes

We identify six empirical foci related to watershed outcomes, grouped into three categories:

- External Influences and Context
  1. Does the number and type of watershed stakeholders affect real and perceived water quality, as measured by both objective indicators and stakeholder perceptions?
  2. How does watershed flow influence real and perceived water quality, based on specific measures and stakeholder perceptions?
- Informational Networks and Legitimacy
  3. In what ways do stakeholder community characteristics impact real and perceived water quality?
  4. How do collective action beliefs among watershed stakeholders shape real and perceived water quality?
  5. To what extent do perceptions of the legitimacy of the watershed decision-making process influence real and perceived water quality?
- Operational Activities
  6. How do the type and number of operational-level activities conducted by watershed stakeholders affect real and perceived water quality?

These focal areas guide our analysis of the relationships between collaborative processes and watershed outcomes.

3.4 Conclusion

This chapter has described the state of the literature regarding the nature of water resource management in the United States. Collaboration occurs at multiple levels and for multiple purposes. But how do we determine if collaborative structures are effective producers of environmental benefits? Koontz and Thomas (2006) are concerned with exactly this and, more importantly, with how we make valid inferences about the effects of collaboration. Numerous small-*n* case studies and a few large-*n* studies have shown that collaboration increases the *perception* of effectiveness among stakeholders. However, to test actual effectiveness measured as water quality sufficient to meet its designated uses, as well as perceptions and the operational level activities these institutions conduct, a more extensive study is needed to generalize about watershed partnerships. Chapter 4 describes an attempt to do this by sampling and investigating watersheds across the fifty states.

References

Agranoff, Robert. 2007. *Managing within Networks: Adding Value to Public Organizations*. Washington, D.C.: Georgetown University Press.

Agranoff, Robert. 2008. Collaboration for Knowledge: Learning from Public Management Networks. In *Big Ideas in Collaborative Public Management*, edited by Lisa Blomgren Bingham and Rosemary O'Leary, 162–194. Armonk, NY: M.E. Sharpe.

Agranoff, Robert. 2012. *Collaborating to Manage*. Washington, D.C.: Georgetown University Press.

Agranoff, Robert, and McGuire, Michael. 2003. *Collaborative Public Management*. Washington, D.C.: Georgetown University Press.

Ansell, Chris, and Gash, Alison. 2007. Collaborative Governance in Theory and Practice. *Journal of Public Administration Research and Theory*, **18**, 543–571.

Bardach, E. 1998. *Getting Agencies to Work Together: The Practice and Theory of Managerial Craftmanship*. Washington, D.C.: Brookings Institution.

Baumgartner, Frank R., and Jones, Bryan D. 2009. *Agendas and Instability in American Politics*. 2nd edn. Chicago: University of Chicago Press. Originally published 1993.

Bidwell, Ryan D., and Ryan, Clare M. 2006. Collaborative Partnership Design: The Implications of Organizational Affiliation for Watershed Partnerships. *Society and Natural Resources*, **19**, 827–843.

Bingham, Lisa Blomgren, and O'Leary, Rosemary (eds). 2008. *Big Ideas in Collaborative Public Management*. Armonk, NY: M.E. Sharpe.

Blomquist, William. 1992. *Dividing the Waters: Governing Groundwater in Southern California*. Lanham, MA: ICS Press.

Blomquist, William. 2009. Multi-Level Governance and Natural Resource Management: The Challenges of Complexity, Diversity, and Uncertainty. In *Institutions and Sustainability: Political Economy of Agriculture and the Environment — Essays in Honour of Konrad Hagedorn*, edited by Volker Beckmann and Martina Padmanabhan, 109–126. Berlin: Springer.

Campbell, Joseph T., Koontz, Tomas, and Bonnell, Joseph E. 2011. Does Collaboration Promote Grass-Roots Behavior Change? Farmer Adoption of Best Practices in Two Watersheds. *Society and Natural Resources*, **24**, 1127–1141.

Carson, Rachel. 1962. *Silent Spring*. Boston: Houghton Mifflin.

Christian-Smith, Juliet, and Gleick, Peter H. (eds). 2012. *A Twenty-First Century U.S. Water Policy*. New York: Oxford University Press.

Cooper, Terry L., Bryer, Thomas A., and Meek, Jack W. 2008. Outcomes Achieved Through Citizen-Centered Collaborative Public Management. In *Big Ideas in Collaborative Public Management*, edited by Lisa Blomgren Bingham and Rosemary, O'Leary, 211–229. Armonk, NY: M.E. Sharpe.

Daniels, Steven E., and Walker, Gregg B. 2001. *Working Through Environmental Conflict: The Collaborative Learning Approach*. Westport: Praeger Press.

Diaz-Kope, Luisa, and Morris, John C. 2019. *Organizational Motivation for Collaboration: Theory and Evidence*. Lanham, MD: Lexington Press.

Diaz-Kope, Luisa, and Morris, John C. 2022. Why Collaborate? Exploring the Role of Organizational Motivations in Cross-Sector Watershed Collaboration. *Politics & Policy*, **50**(3), 516–539.

Diaz-Kope, Luisa, Krawczyk, Kelly, Morris, John C., and Weber, Sindhu. 2024. Commitment to Place as Motivation for Citizen-Based Watershed Collaboration: The "BIMBY" Effect. *Journal of Public and Nonprofit Affairs*, **10**(2), 221–238.

Dietz, Thomas, Ostrom, Elinor, and Stern, Paul C. 2003. The Struggle to Govern the Commons. *Science*, **302**(December), 1907–1912.

Durant, Robert F., Fiorino, Daniel J., and O'Leary, Rosemary (eds). 2004. *Environmental Governance Reconsidered: Challenges, Choices, and Opportunities*. Cambridge, MA: The MIT Press.

Easton, David. 1953. *The Political System*. New York: Alfred A. Knopf.

Elizabeth River Project. 2012. *Money Point: Fish and Shellfish Approve of Restoration Work*.

EPA. 1996 (June). *Watershed Approach Framework*. Tech. rept. United States Environmental Protection Agency.

EPA. 2008. *Handbook for Developing Watershed Plans to Restore and Protect Our Waters*. U.S. Environmental Protection Agency, Washington, D.C.

Feiock, Richard C., and Scholz, John T. (eds). 2010. *Self-Organizing Federalism: Collaborative Mechanisms to Mitigate Institutional Collective Action Dilemmas*. New York: Cambridge University Press.

Feldman, David L. 2007. *Water Policy for Sustainable Development*. Johns Hopkins University Press.

Feldman, David L. 2008. Barriers to Adaptive Management: Lessons from the Apalachicola-Chattahoochee-Flint Compact. *Society and Natural Resources*, **21**, 512–525.

Floress, Kristin J., Prokopy, Linda S., and Allred, Shorna B. 2011. It's Who You Know: Social Capital Social Networks, and Watershed Groups. *Society and Natural Resources*, **24**, 871–886.

Gazley, Beth. 2008. Intersectoral Collaboration and the Motivation to Collaborate. In *Big Ideas in Collaborative Public Management*, edited by L. B. Bingham and R. O'Leary, 36–54. Armonk, NY: M.E. Sharpe.

Gerlak, Andrea K. 2006. Federalism and U.S. Water Policy: Lessons for the Twenty-First Century. *Publius: The Journal of Federalism*, **36**(2), 231–257.

Gleick, Peter H. 2012. The Water of the United States. In *A Twenty-First Century U.S. Water Policy*, edited by Juliet Christian-Smith and Peter H. Gleick, 3–22. New York: Oxford University Press.

Gray, Barbara. 1989. *Collaborating: Finding Common Ground for Multiparty Problems*. San Francisco: Jossey-Bass.

Hale, Kathleen. 2011. *How Information Matters: Networks and Public Policy Innovation*. Washington, D.C.: Georgetown University Press.

Hardin, Garrett. 1968. The Tragedy of the Commons. *Science*, **162**, 1243–1248.

Hardy, Scott D., and Koontz, Tomas M. 2009. Rules for Collaboration: Institutional Analysis of Group Membership and Levels of Action in Watershed Partnerships. *Policy Studies Journal*, **37**(3), 393–414.

Heikkila, Tanya. 2004. Institutional Boundaries and Common-Pool Resource Management: A Comparative Analysis of Water Management Programs in California. *Journal of Policy Analysis and Management*, **23**(1), 97–117.

Heilman, John G., Johnson, Gerald W., Morris, John C., and O'Toole, Jr., Laurence J. 2024. Water Policy Networks in the United States. In *Networks for Water Policy: A Comparative Perspective*, edited by Hans Bressers, Jr. Laurence J. O'Toole and Jeremy Richardson, 80–109. London: Routledge.

Hoornbeek, John. 2011. *Water Pollution Policies and the American States: Runaway Bureaucracies or Congressional Control?* Albany, NY: State University of New York Press.

Hoornbeek, John, Hansen, Evan, Ringquist, Evan, and Carlson, Robert. 2008. *Implementing Total Maximum Daily Loads: Understanding and Fostering Successful Results*. Tech. rept. Report details incomplete in source.

Hoornbeek, John A. 2012. *Water Pollution Policies and the American States: Runaway Bureaucracies or Congressional Control?* State University of New York Press.

Imperial, Mark. 2005. Using Collaboration as a Governance Strategy: Six Lessons from Watershed Management Programs. *Administration and Society*, **37**(3), 281–320.

Irwin, Elise, and Freeman, Mary. 2002. Proposal for Adaptive Management to Conserve Biotic Integrity in a Regulated Segment of the Tallapoosa River, Alabama, USA. *Conservation Biology*, **16**(5), 1212–1222.

Jacobson, Robert B., and Galat, David L. 2008. Design of a Naturalized Flow Regime—An Example from the Lower Missouri River, USA. *Ecohydrology: Ecosystems, Land and Water Process Interactions, Ecohydrogeomorphology*, **1**, 81–104.

Jordan, Jeffrey, and Wolf, Aaron (eds). 2006. *Interstate Water Allocation in Alabama, Florida, and Georgia: New Issues, New Methods, New Models*. Gainesville, FL: The University of Florida Press.

Kahneman, Daniel. 2011. *Thinking, Fast and Slow*. New York: Farrar, Straus and Giroux.

Kash, Jeffrey P. 2008. Enemies to Allies: The Role of Policy-Design Adaptation in Facilitating a Farmer–Environmentalist Alliance. *Policy Studies Journal*, **36**(1), 39–60.

Kettl, Donald F. 2009. The Key to Networked Government. In *Unlocking the Power of Networks: Keys to High-Performance Government*, edited by S. Goldsmith and D. Kettl, 1–14. Washington, D.C.: Georgetown University Press.

King, A. J., Ward, K. A., O'Connor, P., Green, D., Tonkin, Z., and Mahoney, J. 2010. Adaptive Management of an Environmental Watering Event to Enhance Native Fish Spawning and Recruitment. *Freshwater Biology*, **55**, 17–31.

Kingdon, John W. 1995. *Agendas, Alternatives, and Public Policies*. Boston: Little, Brown, and Company. Originally published 1984.

Klijn, Erik-Hans. 1999. Policy Networks: An Overview. In *Managing Complex Networks: Strategies for the Public Sector*, edited by W. J. M. Kickert, E.-H. Klijn and J. F. M. Koppenjan, 14–34. London, U.K.: Sage Publications.

Koehler, Brandi, and Koontz, Thomas M. 2008. Citizen Participation in Collaborative Watershed Partnerships. *Environmental Management*, **41**, 143–154.

Koontz, Thomas M., and Thomas, Craig. 2006. What Do We Know and Need to Know about the Environmental Outcomes of Collaborative Management? *Public Administration Review*, **66**, 111–121.

Leach, William, Pelky, Neil W., and Sabatier, Paul A. 2002. Stakeholder Partnerships as Collaborative Policymaking; Evaluation Criteria Applied to Watershed Management in California and Washington. *Journal of Policy Analysis and Management*, **21**(4), 645–670.

Leahy, Jessica, and Anderson, Dorothy. 2010. "Cooperation Gets It Done": Social Capital in Natural Resource Management along the Kaskaskia River. *Society and Natural Resources*, **23**, 224–239.

Lindblom, Charles E. 1959. The Science of "Muddling Through." *Public Administration Review*, **19**(2), 79–88.

Lubell, Mark. 2003. Collaborative Institutions, Belief Systems, and Perceived Policy Effectiveness. *Political Research Quarterly*, **56**(3), 309–323.

Lubell, Mark. 2004. Collaborative Watershed Management: A View from the Grassroots. *Policy Studies Journal*, **32**, 341–361.

Lubell, Mark. 2005. Do Watershed Partnerships Enhance Beliefs Conducive to Collective Action. In *Swimming Upstream: Collaborative Approaches to Watershed Management*, edited by Paul A. Sabatier, Will Focht, Mark Lubell, Zev Trachtenberg, Arnold Vedlitz and Marty Matlock, 201–232. Cambridge, MA: The MIT Press.

Lubell, Mark. 2007. Familiarity Breeds Trust: Collective Action in a Policy Domain. *The Journal of Politics*, **69**(1), 237–250.

Lubell, Mark, Schneider, Mark, Scholz, John T., and Mete, Mihriye. 2002. Watershed Partnerships and the Emergence of Collective Action Institutions. *American Journal of Political Science*, **46**(1), 148–163.

Mahler, Julianne. 1987. Structured Decision Making in Public Organizations. *Public Administration Review*, **47**(4), 336–342.

Mandarano, Lynn. 2008. Evaluating Collaborative Environmental Planning Outputs and Outcomes: Restoring and Protecting Habitat and the New York-New Jersey Harbor Estuary Program. *Journal of Planning Education and Research*, **27**, 456–468.

McNamara, Madeleine, and Morris, John C. 2021. *Multiorganizational Arrangements for Watershed Protection: Working Better Together*. New York: Routledge.

McNamara, Madeleine W. 2012. Struggling to Untangle the Web of Cooperation, Coordination, and Collaboration: A Framework for Public Managers. *International Journal of Public Administration*, **35**(6), 389–401.

McNamara, Madeleine W. 2016a. Collaborative Management and Leadership: A Skill Set for the Entrepreneur. In *Advancing Collaboration Theory: Models, Typologies, and Evidence*, edited by John C. Morris and Katrina Miller-Stevens, 116–132. New York: Routledge.

McNamara, Madeleine W. 2016b. Unraveling the Characteristics of Mandated Collaboration. In *Advancing Collaboration Theory: Models, Typologies, and Evidence*, edited by John C. Morris and Katrina Miller-Stevens, 65–86. New York: Routledge.

McNamara, Madeleine W., Leavitt, William M., and Morris, John C. 2010. Multiple Sector Partnerships and the Engagement of Citizens in Social Marketing Campaigns. *Virginia Social Science Journal*, **45**, 1–20.

McNamara, Madeleine W., Miller-Stevens, Katrina, and Morris, John C. 2020. Exploring the Determinants of Collaboration Failure. *International Journal of Public Administration*, **43**(1), 49–59.

Menzel, Donald C. 1987. An Interorganizational Approach to Policy Implementation. *Public Administration Quarterly*, **11**(1), 3–16.

Moore, Elizabeth, and Koontz, Tomas. 2003. A Typology of Collaborative Watershed Groups: Citizen-Based, Agency-Based, and Mixed Partnerships. *Society and Natural Resources*, **16**, 451–460.

Moore, Mark H. 2009. Networked Government: Survey of Rationales, Forms, and Techniques. In *Unlocking the Power of Networks: Keys to High-Performance Government*, edited by S. Goldsmith and D. Kettl, 190–228. Washington, D.C.: Georgetown University Press.

Morris, John C., Gibson, William A., Leavitt, William M., and Jones, Shana C. 2013. *The Case for Grassroots Collaboration: Social Capital and Ecosystem Restoration at the Local Level*. Lanham, MD: Lexington Books.

Morris, John C., Gibson, William A., Leavitt, William M., and Jones, Shana C. 2014. Collaborative Federalism and the Emerging Role of Local Nonprofits in Water Quality Implementation. *Publius: The Journal of Federalism*, **44**(3), 499–518.

Olsen, Mancur. 1971. *The Logic of Collective Action: Public Goods and the Theory of Groups*. Cambridge, MA: Harvard University Press.

Ostrom, Elinor. 1990. *Governing the Commons: The Evolution of Institutions for Collective Action*. Cambridge, MA: Harvard University Press.

Ostrom, Elinor. 1998. A Behavioral Approach to the Rational Choice Theory of Collective Action. *American Political Science Review*, **92**(1), 1–22.

Ostrom, Elinor. 1999. Institutional Rational Choice: An Assessment of the Analysis and Development Framework. In: Sabatier, Paul A. (ed), *Theories of the Policy Process*. Boulder, CO: Westview Press.

Ostrom, Elinor. 2005. *Understanding Institutional Diversity*. Princeton, NJ: Princeton University Press.

Ostrom, Elinor, Gardner, Roy, and Walker, James. 1994. *Rules, Games, and Common-Pool Resources*. Ann Arbor, MI: The University of Michigan Press.

Ostrom, Vincent. 1962a. The Political Economy of Water Development. *The American Economic Review*, **52**(3), 450–458.

Ostrom, Vincent. 1962b. The Water Economy and Its Organization. *Natural Resources Journal*, **2**, 55–73.

Ostrom, Vincent. 1974. *The Intellectual Crisis in American Public Administration*. Tuscaloosa, Ala.: Alabama University Press.

Ostrom, Vincent, Tiebout, Charles M., and Warren, Robert. 1961. The Organization of Government in Metropolitan Areas: A Theoretical Inquiry. *American Political Science Review*, **55**, 831–842.

O'Toole, L. J., Jr., Hanf, K. I., and Hupe, P. L. 1997. Managing Implementation Processes in Networks. In *Managing Complex Networks: Strategies for the Public Sector*, edited by Walter J. M. Kickert, Erik-Hans Klijn and Joop F. M. Koppenjan, 137–151. London: Sage Publications.

Poff, N. LeRoy, Allan, David, Bain, Mark B., Karr, James R., Prestegaard, Karen L., Richter, Brian D., Sparks, Richard E., and Stromberg, Julie C. 1997. The Natural Flow Regime: A Paradigm for River Conservation and Restoration. *BioScience*, **47**(11), 769–784.

Putnam, Robert D. 2000. *Bowling Alone: The Collapse and Revival of American Community*. New York: Simon & Schuster.

Rabe, Barry G. 2009. Governing the Climate from Sacramento. In *Unlocking the Power of Networks: Keys to High-Performance Government*, edited by S. Goldsmith and D. Kettl, 62–94. Washington, D.C.: Georgetown University Press.

Repetto, Robert (ed). 2006. *Punctuated Equilibrium and the Dynamics of U.S. Environmental Policy*. New Haven: Yale University Press.

Roe, E. 2001. Varieties of Issue Incompleteness and Coordination: An Example from Ecosystem Management. *Policy Sciences*, **34**, 111–133.

Sabatier, Paul A. 1988. An Advocacy Coalition of Policy Change and the Role of Policy-Oriented Learning Therein. *Policy Sciences*, **21**(2/3), 129–168.

Sabatier, Paul A., and Jenkins-Smith, Hank. 1993. *Policy Change and Learning: An Advocacy Coalition Approach*. Boulder, CO: Westview Press.

Sabatier, Paul A., Focht, Will, Lubell, Mark, Trachtenberg, Zev, Vedlitz, Arnold, and Matlock, Marty (eds). 2005. *Swimming Upstream: Collaborative Approaches to Watershed Management*. Cambridge, MA: The MIT Press.

Scharpf, Fritz W. 1997. *Games Real Actors Play: Actor-centered Institutionalism in Policy Research*. Boulder, CO: Westview Press.

Scheberle, Denise. 2004. *Federalism and Environmental Policy: Trust and The Politics of Implementation*. 2nd edn. Washington, D.C.: Georgetown University Press.

Schlager, Edella. 2002. Rationality, Cooperation, and Common Pool Resources. *The American Behavioral Scientist*, **45**(5), 801–819.

Schlager, Edella, and Blomquist, William. 2008. *Embracing Watershed Politics*. Boulder, CO: University Press of Colorado.

Scholz, John T., and Stiftel, Bruce (eds). 2005. *Adaptive Governance and Water Conflict: New Institutions for Collaborative Planning*. Washington, D.C.: RFF Press.

Scholz, John T., Berardo, Ramiro, and Kile, Brad. 2008. Do Networks Solve Collective Action Problems? Credibility, Search, and Collaboration. *The Journal of Politics*, **70**(2), 393–406.

Shenton, Will, Hart, Barry T., and Chan, Terence U. 2011. Bayesian Network Models for Environmental Flow Decision Making: Latrobe River Australia. *River Research and Applications*, **27**(3), 283–296.

Simon, Herbert. 1947. *Administrative Behavior: A Study of Decision-Making Processes in Administrative Organization*. New York: The Macmillan Company.

Stewart-Koster, B., Bunn, S. E., Mackay, S. J., Poff, N. L., Naiman, R. J., and Lake, P. S. 2010. The Use of Bayesian Network to Guide Investments in Flow and Catchment Restoration for Impaired River Ecosystems. *Freshwater Biology*, **55**, 243–260.

Thomson, Anne Marie, and Perry, James L. 2006. Collaboration Processes: Inside the Black Box. *Public Administration Review*, **66**(Special Issue), 20–32.

Thomson, Anne Marie, Perry, James L., and Miller, Theodore K. 2007. Conceptualizing and Measuring Collaboration. *Journal of Public Administration Research and Theory*, **19**, 23–56.

Walker, Gregg B., Daniels, Steven E., and Emborg, Jens. 2008. Tackling the Tangle of Environmental Conflict: Complexity, Controversy, and Collaborative Learning. *Emergence: Complexity and Organization*, **10**(4), 17–27.

Webber, David J. 2008. Earth Day and Its Precursors: Continuity and Change in the Evolution of Midtwentieth-Century U.S. Environmental Policy. *Review of Policy Research*, **25**(4), 313–332.

Weber, Edward P. 2003. *Bringing Society Back in: Grassroots Ecosystem Management, Accountability, and Sustainable Communities*. Cambridge, MA: The MIT Press.

Weible, Christopher M. 2006. An Advocacy Coalition Framework Approach to Stakeholder Analysis: Understanding the Policy Context of California Marine Protected Area Policy. *Journal of Public Administration Research and Theory*, **17**(1), 95–117.

Williams, Christopher M., Merriman, Connie, and Morris, John C. 2016. A Life-Cycle Model of Collaboration. In *Advancing Collaboration Theory: Models, Typologies, and Evidence*, edited by John C. Morris and Katrina Miller-Stevens, 175–196. New York: Routledge.

Wood, Donna, and Gray, Barbara. 1991. Towards a Comprehensive Theory of Collaboration. *Journal of Applied Behavioral Science*, **27**, 139–162.

4

Findings Revealed: The Impact of Watershed Collaborators on Outcomes

As noted in previous chapters, water quality is intertwined with public health, environmental and ecological sustainability, energy production and generation, and economic development. While a myriad of state and federal laws, actors, and regulations shape the management of the nation's water, it is the Clean Water Act (CWA) that serves as the nation's cornerstone piece of legislation for water quality. Under the CWA, states must take steps to evaluate, protect and, when necessary, restore their waters. To do so, the CWA includes a route for states to develop action plans and pollution limits known as Total Maximum Daily Loads (TMDLs) that are designed to maintain water quality and to meet uses identified by the state. As a voluntary framework, TMDLs offer a pathway for water quality management, yet, as noted in Chapters 2 and 3, progress across the United States in protecting and restoring the nation's waters has proven difficult.

Throughout the United States, watershed collaborations have emerged as complementary or as an alternative to the more state-led approach outlined in the CWA. Often comprising a range of stakeholders, including local governments, nonprofits, agricultural producers, and community members, they have engaged in a multitude of activities and have generated countless "outputs" that are intended to improve water quality. Activities or administrative outputs range from education and awareness campaigns to restoration and testing to implementing BMPs. These outputs, in turn, can work to mitigate or prevent nonpoint source pollution, and thus improve watershed health and quality. Despite the growing recognition and use of watershed collaboration, there are no large-*n* studies that link the revealed activities or outputs of watershed collaborations with revealed water quality outcomes. This chapter is designed to fill that gap.

4.1 Watersheds across the United States

Watersheds within the United States are divided and subdivided into successively smaller hydrologic units: regions, subregions, accounting units, and cataloging units (USGS 2012). In this chapter, we examine revealed outputs and outcomes watersheds at the cataloging unit level, which is considered "a geographic area representing all or part of a surface drainage basin, a combination of drainage basins, or a distinct hydrologic feature" (USGS 2012). Importantly, this level is also commonly used by federal and state agencies and water quality managers to conceptualize problems and craft solutions (Hoornbeek 2012). Figure 4.1 displays one of these cataloging units – specifically, the Lower Tallapoosa watershed (HUC-8 03150110) – and the state in which it is located, including major cities and stream networks. This inset provides a visual anchor for how a single cataloging unit fits within its state context and internal hydrologic structure. There are 2,256 cataloging units across the country classified by 8-digit hydrologic unit codes (HUCs); we typically call them watersheds (USGS 2012). Within each of these watersheds, it is important to remember there is the possibility for several types of waterbodies, such as lakes, rivers, stream segments, and/or estuaries, depending on cataloging unit boundaries. The USGS provides a full list of watersheds on its water resources website.[1]

4.1.1 Sampling Strategy

To assess the utility of collaboration, watersheds from across the 50 states were considered the study population and, because this information is available for all 2,256 watersheds, they also formed our sampling frame. A systematic sampling technique was then applied with a random start, which generated a nearly random sample (Singleton and Straits 2010). With this technique, there is often a periodicity problem when the population is not randomly distributed. In this instance, however, the sampling frame is organized north to south and east to west along hydrographic basins across the fifty states, so this approach should not be problematic. To account for any possibility of this, as well as to have a large enough sample to ensure reliable maximum likelihood estimation (King 1989; Long 1997), considering the reality that complete information may not be available for all watersheds, a random start and sampling interval of four provides 564 observations in the overall sample used for analysis. All 48 of the contiguous states are included in the usable sample. Many watersheds in the sample are contained within a single state, while others cross up to four state boundaries. For example, California is the primary state for 20 watersheds, meaning that the majority of the watershed's waterbodies are contained within the primary state. To summarize the sample, 214 watersheds (69.93%) overlay a single state; 74 watersheds (24.18%) overlay two states; 16 watersheds (5.23%) overlay three states; and two watersheds (0.65%)

[1] United States Geological Survey, "Watershed Boundary Dataset," <http://water.usgs.gov/GIS/wbd_huc8.pdf>.

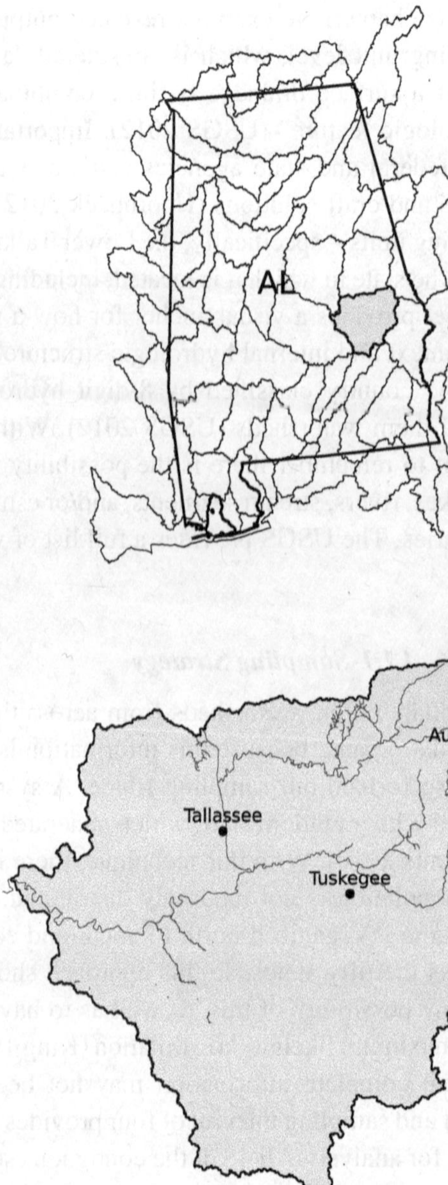

Figure 4.1 Inset maps of the Lower Tallapoosa watershed in Alabama.

overlay four states. Figure 4.2 presents a map of all cataloging units in the contiguous United States. Sampled watersheds used in this study are highlighted in a darker shade.

Any attempt to measure the associations among institutional outputs, outcomes, and water quality levels achieved by any stakeholders is fraught with difficulties.

Figure 4.2 Sampled HUC-8 watersheds in the contiguous United States.

Difficulties include, most notably, external and stochastic shocks, such as wildfires, mine drainage disasters, oil spills, chemical spills, flooding, and droughts. Ostrom (2005, 14) suggests we treat these events as exogenous and fixed, random enough to be captured by error terms in a regression equation. Other variables external to the socioecological processes are not fixed. Across watersheds, they have important and varied influences for which one must control in any attempt for econometric or descriptive analysis of organizational and stakeholder influence on watershed outcomes.

4.2 Revealed Outputs: Watershed Collaboration and Water Quality Outcomes

The EPA's "How's My Waterway" website lists citizen-based groups involved in watershed-related activities at the watershed level. These groups indicated their involvement with the EPA's Adopt a Watershed program at some point over the past two decades. With this information, we created a database of citizen-based group activity within watershed cataloging units that were also included in our sample. We then conducted internet searches for each organization by typing its name into the Google search engine to ensure it is the same organization listed in the database and to verify that it is conducting activities in the cataloging unit. Each of these organizations was then categorized based on their level of activity – local, state, regional, national, or tribal – and affiliation – governmental or nongovernmental, as shown in Box 4.1:

Box 4.1

Organization types represented in the watershed partnerships included in the study. Each partnership could include multiple types.

Local nongovernmental
Local governmental
ENR-related special district
State nongovernmental
State governmental
Regional nongovernmental
Regional governmental
National nongovernmental
National governmental
Tribal governmental
University-affiliated
Federated organization

From this process, we generated 2,795 organizations from across the country operating at multiple levels. It is worth noting that the total list organized by watershed contained 4,261 organizations. Six hundred and thirty survey invitations were distributed in the first round, resulting in 227 completions, a response rate of 36 percent. For the second round, 842 distributed invitations resulted in 305 completions, a response rate of 36 percent. This results in a total of 1,427 distributed surveys and 532 completed and usable responses. Remaining organizations that were not contacted in these two surveys are those that operate at the state, national, and international levels. Because these organizations are involved in so many watersheds, it is most likely difficult for them to respond to inquiries about a specific watershed with the survey instrument employed for this study.

4.3 Revealed Outcomes: Water Quality

To measure the revealed water quality of watersheds, we utilized TMDL classifications considering waterbody designations in 2010. There are several TMDL classifications, as indicated in Table 4.1 and they form the foundation of how we measure revealed water quality (the dependent variable for this chapter). We should note that TMDL classification 5 serves as a measure of the percent of total water bodies within a given watershed that were classified as insufficient to meet their designated uses as well as required a TMDL report and associated mitigation plan. In other words, this is the percent of all measured rivers, streams, lakes, estuaries, and other water bodies for which a known pollutant(s) was identified and considered detrimental to the water body's designated use(s). The information for these data was collected from the EPA's "Water Quality Assessment and Total

Table 4.1 *EPA classification of U.S. water bodies under Clean Water Act Sections 305(b) and 303(d), reflecting their attainment of designated uses and TMDL status.*

Classification	Description
1	All uses attained (Full water body recovery)
2	Some uses attained (Partial water body recovery)
3	Insufficient data
4a	TMDL report developed and approved by the EPA
4b	TMDL report not needed due to alternative mitigation strategy
4c	TMDL report not required due to insufficient hydrology
5	TMDL report needed due to insufficient water quality

Source: U.S. Environmental Protection Agency (EPA).

Maximum Daily Loads Information" website, the EPA's "AskWATERS" website, and state environmental department and agency websites.

Measuring a dependent water quality variable for each watershed with this TMDL data generates several considerations. The first is the view that the TMDL classifications accurately reflect actual water quality conditions in the sample of watersheds. We proceed as though it does, although states do have distinct approaches to their TMDL development, implementation, and enforcement activities, and at the state level, there often exists a rotating emphasis on watersheds within their jurisdiction, incrementally moving from region to region. With this in mind, the data show that classifications 4b and 4c occur with relative infrequency and indicate that a TMDL report is not needed, either because an alternative strategy is in place (4b) or because the water flow is insufficient for a TMDL approach (4c), so they are not very useful for comparisons between watersheds. And because classification 3 indicates insufficient information, one cannot presume these water bodies have sufficient or insufficient water quality for their designated use(s). Additionally, since water quality is determined by each water body achieving its designated use(s), and since designated use(s) often vary across the states as a stream, river, or lake flows from one political boundary to the next, the distinction between TMDL classification 1 and 2, indicating that all or some designated uses have been met, respectively, may be arbitrary, especially for those watersheds that cross state lines. This leaves TMDL classification 5 – a measure of insufficient water quality – for use as a dependent variable. Our measure of water quality, therefore, is the percentage of waterbodies within a given watershed with a TMDL classification 5, indicating a water body that has insufficient water quality to meet its designated use(s) with no remediation plan to meet that use.

Utilizing this as a proxy for revealed water quality involves considering the total number of river miles, lake acres, and estuary square miles and related TMDL classifications of each waterbody in each watershed in the sample. To do so, we gathered the total number of measured miles, acres, and square miles and calculated a percentage of each at TMDL classification 5. To standardize the measure, we summed the percentages and divided by 1 if only rivers are measured in the watershed, 2 if rivers and lakes are measured, and 3 if rivers, lakes, and estuaries are measured. The result is the dependent variable, Percent Insufficient Water Bodies (2010). An additional consideration for using TMDL 5 is related to available data for the lag variable which captures TMDL 5 in a similar manner for 2002. Information provided by state databases (and for some states in EPA databases) for 2002 only included data related to waterbodies classified as TMDL 5; It was not until later in the decade that information on all classifications was provided in a unified report.

4.3.1 Independent Variables

As outlined in Chapter 3 and in our explanatory model, scholars have linked water quality to a host of factors. These, along with their operationalization, are displayed in Table 4.2. The first block of variables represents characteristics of the watersheds states. Table 4.2 follows our model and includes variables approximating socio-economic conditions, ecological conditions, and civic conditions, and information communities. We also include state-level variables because state policymakers are at the center of water policy. State officials, for example, designate water uses, enforce federal and state regulations, provide funding for watershed-related activities and groups, and permit consumptive uses (Morris 2022). And, because the case has been made that many collaborative groups have formed to overcome the collective action problem of NPS pollution, it is important to consider the states role in various ways of impact water quality, especially in terms of ecological and civic conditions.

We begin by highlighting state level characteristics. At the state level, Hoornbeek (2011, 159–60) developed a 30-point state NPS policy activism measure based on (1) "[t]he number of statutory approaches used by the state to facilitate considerations of NPS water pollution concerns in state and local decision-making"; (2) "[t]he strength of the states enforceable authorities for NPS water pollution control"; and (3) "[w]hether states provide significant funding for NPS water pollution control efforts, above and beyond matching funds required under the (CWA Section) 319(h) grant program." These three measures – each set to a 10-point scale – are based on data from 1997 to 2002 and are "combined to create an overall summary measure of state NPS policy activism" (Hoornbeek 2011, 160–61). Hoornbeek (2011) writes that this measure of NPS policy activism score better explains state activism in this arena than traditional measures of state policy activism related to responsive theories, capacity-based approaches, and group-based theories of state policymaking. For this analysis, we apply Hoornbeek's (2011) measure to the state with most of the watershed area within its borders and call it Primary State NPS Activism. Having a variable that reflects conditions at the state level when this investigation is focused on the watershed level necessitates a multilevel modeling approach for statistical analysis.

Water rights doctrine is also important to consider as part of the watershed ecological conditions. Generally, states east of the Missouri River grant water rights based on riparian rights, which originates in English Common Law. Western states typically grant water rights based on a doctrine of prior appropriation. Table 4.3 illustrates which law is applicable in which states. This variable is operationalized as dichotomous for statistical analyses, with a 0 indicating riparian rights and a 1 indicating Prior Appropriation Doctrine. California and Texas, with

Table 4.2 *Variables used in the watershed participation and outcomes analysis, grouped by category. Includes dependent measures, contextual indicators, ecological characteristics, institutional factors, and stakeholder involvement metrics.*

Variable	Type	Description
Dependent variables: Watershed outcomes		
Percent Insufficient Waterbodies (2010)	Continuous*	Percentage of measured units in the watershed with insufficient water quality (TMDL classification 5) as of 2010.
Context		
Socioeconomic Status	Continuous*	Percent of households in the watershed with annual incomes ≥$75,000.
Percent White Population	Continuous*	Percent of population identifying as white.
Population Density	Continuous	Persons per square mile.
Ecological conditions		
Watershed Size	Continuous	Logged square miles of the watershed.
Intermittent Streams	Continuous*	Percent of riverine water with seasonal or intermittent flow.
Institutional and legal factors		
State Water Rights Doctrine	Dichotomous	State water law: 0 = Riparian Rights; 1 = Prior Appropriation.
National Estuary Program (NEP)	Dichotomous	Watershed is part of NEP: 0 = No; 1 = Yes.
Percent Insufficient Waterbodies (2002)	Continuous*	Eight-year lagged version of the 2010 outcome variable.
Percent Watershed in Primary State	Continuous	Percent of watershed land area located in its primary state.
Information community		
Nongovernmental Stakeholders	Count	Number of nongovernmental stakeholder types (see Table 1.2).
Governmental Stakeholders	Count	Number of governmental stakeholder types (see Table 1.2).
State-Level variables		
NPS Pollution Activism	Scalar	Index (0–30) combining statutory support, enforcement capacity, and funding for nonpoint source (NPS) pollution control.

*Indicator normalized and scaled for statistical modeling.
Source: Montana State University "Statistics Group Hydrologic Unit Project."

a dual doctrine, are coded as 1 to capture the riparian doctrine that applies to surface waters.

Turning to watershed-level covariates, much of the context data captures socioeconomic and ecological conditions and is collected from the National Oceanic and

Table 4.3 *Classification of US states by dominant water rights doctrine: Riparian Rights vs. Prior Appropriation.*

Riparian Rights		Prior Appropriation	
Alabama	Arkansas	Alaska	Arizona
Connecticut	Delaware	California	Colorado
Florida	Georgia	Hawaii	Idaho
Illinois	Indiana	Kansas	Montana
Iowa	Kentucky	Nebraska	Nevada
Louisiana	Maine	New Mexico	North Dakota
Maryland	Massachusetts	Oklahoma	Oregon
Michigan	Minnesota	South Dakota	Texas
Mississippi	Missouri	Utah	Washington
New Hampshire	New Jersey	Wyoming	
New York	North Carolina		
Ohio	Pennsylvania		
Rhode Island	South Carolina		
Tennessee	Vermont		
Virginia	West Virginia		
Wisconsin			

Note: California and Texas are *Dual Doctrine* states, where both Riparian Rights and Prior Appropriation coexist.

Atmospheric Administration (NOAA). Data from the 2,000 census reflects conditions in the watershed 10 years before the dependent variable measuring water quality in 2010 and offers a baseline. This historical measure also captures characteristics of the watershed that may impact water quality as measured in 2010. The following variables are gathered from this website: socioeconomic status, measured as the percentage of households within watershed boundaries with annual incomes of $75,000 or more; percent white population, measured as the percent of the population that is white; population density, measured as the number of persons per square mile; and watershed size, measured as the number of square miles of the geographic area covered by the watershed.

We also include explicit measures of the watershed's ecological conditions. A measure of the percent of the water flowing through the intermittent watershed – the percent that flows seasonally – is gathered to account for flow variability from Montana State University's Statistics Group Hydrological Unit Project. While intermittent flow does not capture all the important aspects of water flow, it is still fundamental to the measure (Poff et al. 1997) as it captures precipitation, snowpack melt, and the watershed's ability to assimilate pollutants. Two additional contextual control variables are gathered from EPA websites. The first is a dichotomous measure of whether the watershed is part of the National

Estuary Program (NEP). This is collected from the "Surf Your Watershed" website and is coded 1 if the watershed is involved in the program. Due to the nature of the program, it often spans multiple watersheds and brings stakeholders together for collaborative decision-making processes related to the governance of an estuary. Secondly, from the "Water Quality Assessment and Total Maximum Daily Loads Information" website and the EPA's "AskWATERS" website, a measure of percent insufficient water bodies (2002) is an eight-year time lag variable of the dependent variable of the percent insufficient water bodies in 2010. This measure accounts for the problem severity that watershed stakeholders faced eight years before the 2010 measure used in this analysis.

4.4 Analysis

4.4.1 Univariate Results

Table 4.4 shows the summary statistics for all the variables associated with Table 4.2. It is worth mentioning at the outset that an average of 41% ($\hat{\sigma} = .30$) of the watersheds in the sample have majority insufficient water quality, which is a statistically significant difference from eight years prior when the figure was 49% ($\hat{\sigma} = 0.32$) of the sample were insufficient ($\hat{\rho} = 0.39; p < 0.001$). While this is impressive, further study is required to determine causal factors leading to this improvement. Moving to organizations involved in the surveyed watersheds, Table 4.4 illustrates that an average of 0.96 ($\hat{\sigma} = 3.07$) local-level government agencies are involved. However, with such a wide range, zero to 32, this mean may be an inadequate representation of central tendency, possibly due to the amount of federally owned lands and waters in which no local-level authority is involved. Examples of these local organizations include Clackamas County Water Environment Services, Frederick County Forestry Board, and the Washington County Redevelopment Authority.

The average number of local-level nongovernmental organizations participating in watershed-related activities is 3.20 ($\hat{\sigma} = 5.16$), although this greatly varies from watershed to watershed, ranging from 0 to 40. Examples of these include Friends of the Savannah River Basin, Cadron Creek Alliance, and the Soque River Watershed Association. At a unique level (Mullin 2009), an average 0.49 ($\hat{\sigma} = 1.31$) natural resource-related special districts are actively involved in watershed related activities. Examples of these include the Nez Perce Soil and Water Conservation District, Cedar Bayou Navigation District, and the Upper Salinas Las Tablas Resource Conservation District. Federated organizations, such as the Washington County chapter of the Izaak Walton League of America, the North Jersey Sierra Club, and the Alabama affiliate of the National Wildlife Federation have a unique

Table 4.4 *Summary statistics for stakeholder participation and contextual variables across 306 watershed collaborations. Measures include stakeholder types, demographic indicators, and legal-institutional controls.*

Variable	Mean	SD	Min	Max
Dependent variable				
Percent Insufficient Waterbodies 2010 (DV)	0.41	0.30	0	1
Independent variables (Watershed level, $n = 306$)				
Local-level NGOs	3.20	5.16	0	40
Local-level government agencies	0.96	3.07	0	26
Natural resource-related special districts	0.49	1.31	0	9
State-level NGOs	1.57	2.03	0	12
State-level government agencies	1.09	1.84	0	11
Regional-level NGOs	0.52	0.99	0	6
Regional-level government organizations	0.07	0.29	0	2
National-level NGOs	0.11	0.39	0	3
National-level government organizations	0.69	1.55	0	9
University-affiliated organizations	0.51	1.01	0	7
Federated organizations	0.68	1.04	0	5
Tribal government organizations	0.07	0.41	0	4
Control variables				
National Estuary Program (NEP) Watershed	0.23	0.42	0	1
Socioeconomic Status (2000)	0.15	0.07	0.04	0.50
Population Density (2000)	154.62	436.24	0.13	5334.05
Percent White Population (2000)	0.85	0.14	0.09	0.99
Watershed Size (Square Miles)	1477.72	880.02	73.40	8574.09
Percent Intermittent Stream Flow	43.48	27.06	0	98.48
Percent Watershed in Primary State	0.93	0.14	0.44	1.00
Prior Appropriation Water Doctrine (Primary State)	0.45	0.49	0	1
Percent Insufficient Waterbodies 2002 (DV lag)	0.49	0.32	0	1
NPS Pollution Activism (Primary State [1997–2002])	12.39	7.73	2	25.5

presence in these policy subsystems, operating at the local level with influences, resources, and capacity provided from their national or international parent organizations. These organizations range from 0 to 5 participating at the watershed-level, with an average of 0.68 ($\hat{\sigma} = 1.04$) in each watershed.

Turning to state and regional levels, there is also wide variation in the number of organizations involved at the watershed level. An average 1.57 ($\hat{\sigma} = 2.03$) state-level nongovernmental organizations, ranging from 0 to 12, are participating in watershed-related activities. These include organizations such as the Alabama Clean Water Partnership, the Kansas Association of Wetlands and Streams, and the

Texas Riparian Association. An average 1.09 ($\hat{\sigma} = 1.84$) state-level government agencies and departments are participating in watershed-related activities. Examples of these include the Alaska Department of Natural Resources, the Connecticut Department of Energy and Environmental Protection, and the New Hampshire Natural Heritage Bureau. An average 0.52 ($\hat{\sigma} = 0.99$) regional-level nongovernmental organizations are active in watershed-related activities, ranging from 0 to 6. These include organizations such as the Rocky Mountain Environmental Associates, the Greater Yellowstone Coalition, and the Coalition for the Delaware River Watershed. The number of regional governmental organizations engaged at the watershed level ranges from 0 to 2, with an average of 0.07 ($\hat{\sigma} = 0.29$), with examples such as the Great Lakes Restoration Initiative, the Northeast Ohio Areawide Coordinating Agency, and the Lower Connecticut River Valley Council of Governments.

Watersheds have an average of 0.11 ($\hat{\sigma} = 0.39$) national-level nongovernmental organizations participating in watershed-related activities. This number ranges from 0 to 3 and includes organizations such as American Rivers, Save our Streams, and the Coastal Conservation Association. National-level government agencies and organizations are active in an average 0.69 ($\hat{\sigma} = 1.55$) watersheds. This number ranges from 0 to 9 and includes organizations such as the National Park Service, National Wildlife Refuges, and the US Army Corps of Engineers. University-affiliated organizations, such as the Watershed Health Clinic housed within the University of Montana and the Watershed Institute of Central Virginia, are active in an average 0.51 ($\hat{\sigma} = 1.04$) watersheds. This number ranges from 0 to 7 and often involves university extension offices. Finally, tribal government organizations are involved in an average 0.07 ($\hat{\sigma} = 0.41$) watersheds. This number ranges from 0 to 4 and includes the Confederated Tribes of Siletz Indians, the Little Traverse Bay Band of Odawa Indians, and the watershed division of The Nez Perce Tribe.

Beyond organizational involvement and the information environment, Table 4.4 describes several other variables that may influence water quality. Among these is a variable for the National Estuary Program (NEP), which makes incumbent on stakeholders a collaborative management approach as it necessarily brings together multiple parties at multiple levels to conduct watershed-related activities in several estuaries. As a dichotomous measure, it shows an average of 0.23 ($\hat{\sigma} = 0.42$) watersheds in this sample are involved in the NEP. Socioeconomic status (SES) is measured as the percentage of households within the watershed with an annual income greater than or equal to \$75,000 in 2000.

Table 4.4 shows that an average of 15% ($\hat{\sigma} = 7\%$) of watershed households have incomes at this level, although this number ranges from 4% to 50%. As for population density, this table shows there were an average 154.62 persons per square mile ($\hat{\sigma} = 436.24$) living in sampled watersheds in 2000; this number ranges from 0.13 to 5334.05. This measure is related to the watershed size variable, which indicates that the average watershed is 1477.72 square miles ($\hat{\sigma} = 880.02$), ranging from 73.40 to 9574.09. Finally, sampled watersheds report an average 85% ($\hat{\sigma} = 14\%$) white population in 2000, with a wide range of 9% to 99%. Apart from the NEP variable, these data were gathered from the 2000 US Decennial Census.

Four additional variables measure ecological conditions and civic conditions. First, percent intermittent stream flow indicates variation in seasonal flows, as well as crudely measuring precipitation and snowpack melt. Sampled watersheds have an average of 43% ($\hat{\sigma} = 27\%$) intermittent flows; this ranges from no intermittent streams to nearly all seasonally intermittent flows (98%). An average of 93% ($\hat{\sigma} = 14\%$) of the geographic area of sampled watersheds is contained within a single state, referred to here and in the table as the primary state; this ranges from 44% to 100%. A dichotomous indicator for the prior appropriation (Western) water doctrine shows that 45% of sampled watersheds use this approach for determining water rights. The nonpoint source (NPS) pollution activism score measures state-level NPS pollution activism over the years 1997–2002; this score is based on three state-level characteristics using a 30-point scale, averaged over the number of states that the watershed overlays. The average NPS pollution activism score for sampled watersheds is 12.26 ($\hat{\sigma} = 7.83$), ranging from 2 to 25.5. Also included as an eight-year lag on the dependent variable is the percent of waterbodies that were designated as insufficient in 2002.

4.4.2 Bivariate Analysis of Water Quality Outcomes and Predictors

Building on our descriptive results, we turn next to unpacking potential bivariate relationships between water quality outcomes in 2010 and key theoretical predictors. We employ Pearson correlations for continuous variables and difference-of-means tests for categorical factors. Given the social and ecological complexity of watersheds and the complicated interactions among actors, land use, and sociopolitical arrangements within watersheds, this section analyzes patterns of association between water quality and various predictors. Following our explanatory model, we address three types of factors: (1) continuous external influences, contextual factors, and binary institutional features (e.g., presence of compacts, agency involvement, and the types of organizations involved in watershed efforts). To do so, we leverage Pearson correlations for continuous variables and difference-of-means tests for categorical factors.

Table 4.5 *Bivariate correlations between continuous predictors and participation dimensions in watershed collaborations.*

Predictor	Pearson *r*
State-level environmental activism	0.02
Socioeconomic status (SES) index	0.13*
Population density (per sq. mi, 2000)	0.21***
Urban population (% Urban, 2000)	0.22***
White population (% White, 2000)	−0.18**
Number of water districts in watershed	0.15**
Total partner organizations involved	−0.04

Note: Significance levels: $^{*}p < 0.10$, $^{**}p < 0.05$, $^{***}p < 0.01$.

Table 4.6 *Model-estimated participation levels across institutional and organizational conditions. Values reflect group means and differences (full sample, n* = 345*). Significance noted via asterisks.*

Factor	Group 1 Mean	Group 2 Mean	Difference
Water rights regime	0.402	0.418	+0.016
Watershed scope	0.416	0.388	−0.028
Interstate compact present	0.413	0.224	−0.189*
NEP designation	0.420	0.368	−0.052
Local government involvement	0.396	0.454	+0.058
State government involvement	0.423	0.388	−0.035
Federal agency involvement	0.419	0.378	−0.041
Stakeholder conflict†	0.392	0.404	+0.012

†Conflict refers to reported stakeholder disagreement or friction.
Note: $^{*}p < 0.05$.

Tables 4.5–4.8 highlight how our various external, contextual, institutional, and collaborative characteristics relate to revealed water quality outcomes. Table 4.5 shows correlations between 2010 water quality outcomes and our continuous predictor variables. Many of these variables represent external characteristics, civic conditions, and stakeholder communities, such as socioeconomic status, urbanization, racial composition, and collaborative capacity. Table 4.6 shifts the focus to state and federal characteristics and compares mean water quality levels across watersheds with and without institutional actors, including water rights, the involvement of more than one state, and federal participation. Finally, Tables 4.7 and 4.8 examine the relationships among organizational types, for example, regional nongovernmental organizations, relative to water quality and with differences in revealed water quality outcomes.

Table 4.7 *Bivariate Pearson correlation coefficients for organization types predicting participation levels (Subset A). Significance indicated via asterisk.*

Organization type	Pearson **r**
Local NGOs	−0.003
Local government	0.069
Special districts	0.007
State NGOs	−0.040
State government	−0.032
Regional NGOs	−0.113*
Regional government	0.054
National NGOs	0.027
National government	−0.044
University-affiliated	−0.055
Tribal government	−0.009
Federated organizations	0.025

Note: Pearson r reflects strength and direction of bivariate relationships. $^{*}p < 0.05$.

Table 4.8 *Mean participation rates in watershed collaborations based on organizational presence (Subset B). Values reflect group means and differences. Asterisk indicates statistical significance at* $p < 0.05$.

Organization type	No presence	Presence	Difference
Local NGOs	0.406	0.403	−0.003
Local government	0.390	0.459	0.069
Special districts	0.403	0.410	0.007
State NGOs	0.430	0.390	−0.040
State government	0.418	0.386	−0.032
Regional NGOs	0.437	0.374	−0.063*
Regional government	0.406	0.447	0.041
National NGOs	0.406	0.420	0.014
National government	0.419	0.378	−0.041
University-affiliated	0.419	0.378	−0.041
Tribal government	0.408	0.402	−0.006
Federated organizations	0.403	0.411	0.008

Note: Difference column shows mean participation rate gaps based on organizational presence. $^{*}p < 0.05$.

4.4.2.1 Bivariate Correlations with Continuous Predictors

We begin by examining Table 4.5 to determine whether any of our continuous factors influence revealed water quality outcomes. Results from the Pearson correlations offer early clues as to which variables are more strongly associated with water quality impairment.

State-level environmental activism and the size of collaborative partnerships do not display significant relationships, highlighting the complexity of institutional influences. Interestingly, the number of specialized water districts is positively associated with better outcomes ($r = +0.15$, $p < .01$), pointing to the possible importance of dedicated management capacity.

4.4.2.2 Bivariate Differences by Categorical Factors

We next compare water quality outcomes based on differing institutional characteristics. These comparisons examine whether specific institutional features, such as water rights regimes, multistate scope, NEP designation, and the involvement of government agencies, are associated with water quality outcomes.

Table 4.6 summarizes mean differences in water quality impairment across key categorical factors. The results indicate that, among the institutional variables considered, only the presence of an interstate compact is significantly associated with water quality outcomes. Watersheds governed by interstate compacts exhibit, on average, 19 percentage points higher levels of impaired water quality compared to those without such compacts ($p < 0.05$). This suggests that complex, multijurisdictional governance arrangements may hinder effective coordination and remediation efforts, potentially due to increased transaction costs or conflicting priorities among member states.

Other institutional features, including water rights regime (riparian vs. prior appropriation), whether the watershed spans multiple states, NEP designation, and the involvement of local, state, or federal agencies, do not show statistically significant differences in water quality outcomes. Similarly, the presence of reported stakeholder conflict is not associated with significant differences in impairment levels. These findings reinforce the notion that not all institutional arrangements or forms of government involvement translate into measurable improvements (or declines) in water quality, and that the challenges posed by interstate coordination may be particularly salient.

4.4.2.3 Bivariate Correlations with Organization Types

Our next set of results explores if the presence of specific organizational types is linked to differences in water quality. Table 4.7 summarizes these relationships by organizational type and scale and ranges from local nongovernmental organizations to the federal/national government.

The bivariate Pearson correlations in Table 4.7 indicate that, among all organization types, only regional NGOs are significantly associated with improved water quality outcomes ($r = -0.113, p = 0.042$). This negative correlation suggests that watersheds with greater participation from regional NGOs tend to have lower levels of impaired water quality. The effect size, while modest, is notable given the lack of significant associations for other organizational types.

Other organization types – including local NGOs, local governments, special districts, state-level organizations, national organizations, university-affiliated groups, tribal governments, and federated organizations – do not show statistically significant relationships with water quality outcomes. For example, local government involvement approaches significance ($p = 0.102$) but does not meet conventional thresholds. The absence of significant associations for most organization types suggests that the scale and scope of organizational engagement may be more important than the mere presence of organizations.

Overall, these results reinforce the unique role of regional NGOs in supporting water quality improvements, likely due to their ability to coordinate across jurisdictions, leverage resources, and implement practices at a broader scale than purely local or state actors. The findings highlight the importance of organizational capacity and regional coordination in achieving measurable environmental outcomes.

4.4.2.4 Mean Differences in Water Quality Outcomes by Organizational Presence

To further explore the dynamics between organizational type and water quality outcomes, we next compare average water quality in watersheds with and without each organizational type. In other words, are certain types of organizations associated with improved water quality? Table 4.8 presents mean water quality impairment rates for watersheds grouped by the presence or absence of each organizational type, along with the mean differences and significance tests.

The results in Table 4.8 reinforce the earlier correlation findings: only regional NGOs are associated with a statistically significant difference in water quality outcomes. Specifically, watersheds with regional NGO involvement exhibit, on average, 6.3 percentage points lower levels of impaired water quality ($p = 0.042$) compared to those without such organizations. This effect is both statistically and substantively meaningful, highlighting the unique contribution of regional NGOs to watershed health.

For other organizational types – including local governments, special districts, state-level organizations, and national entities – no statistically significant differences are observed. While local government presence approaches significance ($p = 0.102$), the effect is modest and does not meet conventional thresholds. These results suggest that it is not simply the presence of organizations, but rather the scale, coordination capacity, and regional reach of NGOs that are most strongly linked to improved water quality outcomes. The findings underscore the importance of organizational structure and capacity in driving tangible environmental improvements, rather than relying solely on the number or diversity of participating groups.

4.4.2.5 Summary of Bivariate Analysis Findings

Taken together, these bivariate analyses reinforce the critical importance of contextual and institutional factors over simple counts of collaborative activity as currently measured. Urbanization and socioeconomic conditions consistently emerge as stronger predictors of water quality outcomes than the sheer presence of stakeholder organizations. Among institutional arrangements, the presence of interstate compacts appears to pose a significant barrier to water quality improvements, highlighting the coordination and governance challenges inherent in multi-state water management.

Most notably, the analysis of organizational types reveals that not all collaborators are equally positioned to affect change. Regional NGOs stand out as the only stakeholder group with a statistically significant positive association with improved water quality. This suggests that the scale and capacity of participating organizations – not just their involvement – are critical determinants of collaborative success. These findings underscore the need to move beyond descriptive accounts of who participates in watershed collaborations to examine how organizational capacity, scope, and governance arrangements interact to shape environmental outcomes.

The following section builds directly on these insights, presenting multivariate models that assess the combined influence of these factors while controlling for potential confounders and testing the robustness of these initial relationships.

4.4.3 Multilevel Results

Watersheds sit at the intersection of overlapping institutions. At the federal level, agencies vary in their involvement, from managing infrastructure in national parks to enforcing endangered species protections. EPA regional offices maintain distinct relationships with state and local actors (Scheberle 2004). Meanwhile, states diverge in legal doctrines, support for watershed partnerships, and the level of policy engagement around pollution. States also differ in how they designate uses for waterbodies and how they staff and fund TMDL development. Finally, as detailed earlier in this volume, diverse stakeholders – governmental and nongovernmental – intervene directly at the watershed scale.

The following section evaluates whether those institutional and organizational differences correspond with measurable improvements in water quality. Using a national dataset, we estimate multilevel models that assess how stakeholder presence, demographic context, and prior water quality relate to 2010 outcomes. This approach enables a more rigorous test of our theoretical framework, allowing us to evaluate whether certain configurations of actors and state environments consistently correspond with stronger environmental performance.

Water quality outcomes are measured as the percentage of waterbodies in each watershed classified as impaired in 2010 under the Clean Water Act. Given the nested structure of the data – watersheds within states – and the relevance of state-level capacity, multilevel modeling is appropriate. It accounts for both within-state variation (e.g., stakeholder involvement) and between-state variation (e.g., policy activism). Following standard practice (Dahill-Brown and Lavery 2012; Miller 2013), we begin with an unconditional model to estimate the intraclass correlation coefficient (ICC). The ICC of 0.293 confirms that nearly 30% of variation in outcomes lies between states, justifying a hierarchical approach.

The multilevel model is specified as follows:

$$\text{Level 1 (within states):} \quad Y_{ij} = \beta_{0j} + \beta_{1j}X_{ij} + r_{ij}$$

$$\text{Level 2 (between states):} \quad \beta_{0j} = \gamma_{00} + \gamma_{01}W_j + u_{0j}$$

$$\beta_{1j} = \gamma_{10} + \gamma_{11}W_j + u_{1j}$$

where

Y_{ij} is the proportion of impaired waters in watershed i, nested within state j.
X_{ij} are watershed-level predictors (e.g., number of regional NGOs and demographics).
W_j are state-level predictors (e.g., NPS policy activism).
r_{ij}, u_{0j}, u_{1j} are the residuals at each level.

This structure allows intercepts and slopes to vary by state, capturing cross-level interactions between state institutional environments and local stakeholder configurations.

4.4.3.1 Model Results

Five models are estimated and presented in Table 4.9. Each builds incrementally to assess how institutional, organizational, and demographic factors shape water quality outcomes. Model 1 uses a fixed-effects structure with no state-level terms, serving as a baseline. Model 2 introduces a mixed-effects structure to account for variation across states. Model 3 adds a state-level measure of NPS policy activism. Model 4 is a reduced-form specification that retains only statistically significant predictors. Model 5 combines this reduced model with the activism variable.

Across all models, regional NGOs stand out. Their presence is consistently associated with improved water quality. Specifically, each additional regional NGO corresponds to a 4–5% reduction in the proportion of impaired waters ($p < 0.01$). Prior water quality is also a strong and stable predictor: watersheds impaired in 2002 are significantly more likely to remain impaired in 2010, with coefficients consistently above 0.34 ($p < 0.001$). Demographics matter as well.

Table 4.9 *Multilevel regression models estimating effects on insufficient water quality (2010). Five models compare fixed vs. mixed state effects and full vs. fit model specifications. Coefficients ($\hat{\beta}$) with standard errors ($\hat{\epsilon}$) in parentheses.*

	Full Model 1 State fixed effects		**Full Model 2** State mixed effects		**Full Model 3** State factors mixed		**Fit Model 4** Mixed effects		**Fit Model 5** Factors mixed	
Watershed-level effects										
Local NGOs	0.004	(0.005)	0.004	(0.007)	0.004	(0.007)				
Local gov't org's	0.005	(0.007)	0.005	(0.007)	0.005	(0.007)				
Special districts	0.007	(0.016)	−0.004	(0.015)	−0.003	(0.015)				
State NGOs	−0.010	(0.010)	−0.001	(0.010)	0.006	(0.010)				
State gov't org's	0.003	(0.015)	0.010	(0.015)	0.010	(0.015)				
Regional NGOs	−0.045**	(0.017)	−0.052**	(0.017)	−0.054**	(0.017)	−0.044**	(0.015)	−0.045**	(0.015)
Regional gov't org's	0.011	(0.052)	0.005	(0.048)	0.010	(0.049)				
National NGOs	0.082	(0.048)	0.080	(0.080)	0.073	(0.045)				
National gov't org's	−0.012	(0.017)	−0.004	(0.016)	−0.003	(0.016)				
University affiliates	−0.029	(0.017)	−0.022	(0.017)	−0.020	(0.017)				
Tribal gov't org's	−0.014	(0.045)	−0.002	(0.041)	0.001	(0.041)				
Federated org's	0.002	(0.020)	−0.020	(0.022)	−0.030	(0.022)				
% insufficient waters 2002	0.361***	(0.048)	0.350***	(0.052)	0.370***	(0.050)	0.349***	(0.052)	0.357***	(0.051)
NEP watershed	−0.066	(0.038)	−0.024	(0.040)	−0.021	(0.038)				
SES (2000)	0.298	(0.252)	0.210	(0.248)	0.137	(0.242)				
% watershed in state	0.101	(0.106)	0.172	(0.099)	0.157	(0.099)				
Population density	0.000	(0.000)	0.000	(0.000)	0.000	(0.000)				
% white pop. (2000)	−0.299**	(0.102)	−0.215*	(0.099)	−0.205**	(0.103)	−0.295**	(0.104)	−0.267**	(0.101)
Watershed size (mi^2)	−0.000	(0.000)	0.000	(0.000)	−0.000	(0.000)				
% intermittent flow	0.001	(0.001)	0.001	(0.001)	0.001	(0.001)				
Prior appropriation	0.043	(0.038)	0.026	(0.054)	0.047	(0.049)				
Intercept	0.320*	(0.150)	0.184	(0.151)	0.177	(0.147)	0.541***	(0.095)	0.495***	(0.091)
State-level effects										
ln($\hat{\sigma}$) within-state	−1.395***	(0.040)	−1.527***	(0.043)	−1.427***	(0.044)	−1.488***	(0.043)	−1.491***	(0.043)
ln($\hat{\sigma}$) between-state			−2.114***	(0.167)	−4.885***	(0.282)	−2.046***	(0.161)	−4.843***	(0.330)
ln($\hat{\sigma}$) NPS activism					−2.906***	(0.745)			−2.604***	(0.480)
Model fit statistics										
BIC	146.34		117.34		119.25		39.98		42.29	
χ^2		127.42***		92.44***		98.07***		60.40***		62.97***

Note: Level 1 observations = 306; Level 2 groups = 45.
$^{*}p < 0.05$, $^{**}p < 0.01$, $^{***}p < 0.001$.

A higher percentage of white residents in 2000 predicts lower impairment rates in 2010 ($p < 0.05$), suggesting the lingering influence of demographic privilege in environmental outcomes.

State-level NPS activism, however, consistently fails to reach statistical significance. This aligns with findings from Morris (2022), reinforcing the idea that formal policy commitment is not, by itself, enough to drive measurable environmental improvement. Absent strong implementation mechanisms or stakeholder engagement, policy ambition appears to have limited traction.

4.4.3.2 Model Comparison

Beysian Information Criterion (BIC) values are used to compare model fit. Model 4 offers the best empirical fit (BIC = 39.98) and distills the analysis to three key predictors: regional NGOs, percent white population, and prior impairment. In this model, each additional regional NGO is associated with a 4.4% reduction in impairment ($p < 0.01$). The 2002 impairment rate remains a robust predictor of 2010 outcomes ($\hat{\beta}ij = 0.349$, $p < 0.001$), and percent white population is negatively associated with impairment ($\hat{\beta}ij = -0.295$, $p < 0.01$). Variance estimates are statistically significant at both levels: within-state variance is $\hat{\sigma}^2 = 0.014$ (SE = 0.005, $p < 0.001$), and between-state variance is $\hat{\sigma}^2 = 0.047$ (SE = 0.004, $p < 0.001$).

Model 2, which retains all covariates but excludes NPS activism, also performs well (BIC = 117.34). It estimates a slightly larger effect for regional NGOs (−5.2, $p < 0.01$), and again finds that percent white population and prior impairment remain significant. The variance components remain statistically distinguishable, reinforcing the need for a multilevel structure.

Models 3 and 5, both of which include the state-level activism variable, do not improve model fit or change the substantive conclusions. Their coefficients lack statistical significance, and their inclusion does not meaningfully affect the interpretation of other predictors. This suggests that while NPS activism may be relevant conceptually, it does not translate into measurable differences in water quality outcomes in this sample.

4.5 Implications

The models presented here demonstrate the socioecological complexity that is inherent to watersheds and insufficient water quality. That said, we begin by observing what the data show as not significant. The emphasis of collaborative natural resource management is often placed at the local level (e.g., the watershed approach and grassroots environmental management). However, our results suggest that the number of organizations working at this level are not associated with sufficient water quality. While we cannot isolate a cause, several possibilities exist. Perhaps it is a lack of institutional capacity at this level, a scalar mismatch, or that transaction costs are too high for local organizations to work effectively with their counterparts without the assistance of actors at different levels. There is also the possibility that (1) local actors favor economic development over environmental protection, or (2) concerned local actors are more likely to form associations as well as partnerships when environmental conditions are overly stressed. To the latter point, severe problems may be more likely to generate a greater number of actors needed to conduct comprehensive restoration, remediation, and educational activities, and the outputs they generate may require longer time horizons to produce beneficial outcomes. It may also be that while local stakeholders are not linked to revealed quality outcomes, they do generate social and capacity benefits from interactions with those organizations located in at higher classifications, for example, in a regional association in a loose network of actors that share information and resources across that network, from best practices to financial assistance programs. The evidence here cannot specify any causal processes with any certainty for this relationship.

Our results highlight the following as potential drivers of water quality outcomes. First, they show that the number of regional-level nongovernmental stakeholder organizations involved in watershed-related activities has a statistically significant negative impact on insufficient water quality. Second, there appears to be a relationship among past water quality outcomes and present-day water quality. Finally, results suggest a dynamic among demographics and water quality outcomes. With these relationships in mind, the key questions become why; and why do they matter for watershed collaboration?

4.5.1 An Enhanced Role for Regional Organizations

We observe that a one-unit increase in the number of regional organizations involved in a watershed collaboration is associated with a reduction in insufficient water quality by about 5%. In short, this relationship suggests that regional collaboration plays a key role in improving water quality outcomes. Regional-level associations or groups offer watershed collaborations several benefits that may support improved water quality. First, as groups that operate across a region (that likely involves multiple local units of governments and nonprofits) they are likely able to contribute to improved coordination across stakeholders, to facilitate stakeholder resource sharing, and to support the implementation of water quality improvement practices across the region. This is an explanation consistent with the voluminous case-based literature on watershed collaboration (see Sabatier et al. 2005; Morris et al. 2013). Additionally, regional organizations may support collaborative efforts by offering diverse expertise, perspectives, or skills such as grant-writing yet still hold local knowledge and connections that are relevant to achieving water quality outcomes. Additionally, organizations operating at a regional level may be able to enhance stakeholder education or legitimacy as well as raise public awareness across an entire region (rather than one city or neighborhood) contributing to greater acceptance or use of best management practices (BMP) and water quality improvement/protection efforts (Genskow and Prokopy 2009; Kneese 2009).

4.5.2 The Importance of Time and Past Practices

Our strongest predictor of insufficient water quality in 2010 was the level of water quality in 2002. In short, watersheds with poor water quality in 2002 were more likely to report poor quality outcomes in 2010. This finding makes intuitive sense. First, many water quality issues are path-dependent, complex, and without sizable restoration efforts can take decades to improve. And, as noted in Chapter 1, federal policies designed to protect water quality have only emerged within the last 50 years and states have only meaningfully engaged with Section 303 for the past 30 or so years. This means that historical pollutants including heavy and radioactive metals, agricultural and municipal runoff, industrial discharges, and other wastes have accumulated in water bodies and nearby soils, contributing to long-term site contamination. These pollutants may persist for decades and cost millions to remediate and/or remove, representing a significant challenge for cash-starved state and local governments. It is also possible that shocks and other events may lead to contaminants being released harming water quality.

As noted, restoring water quality does not happen overnight and, in fact, may take decades. In this sense, understanding the goals of the collaboration is critical. If the goal is protecting existing pristine waters, outputs such as enhancing buffers or implementing other natural filtration systems, may work to prevent additional contamination. However, such efforts may take years to contribute to water quality improvements. Similar dynamics are likely to govern water quality infrastructure and outcomes. Outdated municipal stormwater systems and agricultural drainage infrastructure, contribute to water quality issues but often cannot be changed quickly meaning that water quality restoration will likely be a slow and incremental process (Schaider et al., 2019).

4.5.3 The Possible Role of Demographics?

We also observe that watersheds with a higher percentage of the white residents in 2000 are associated with lower levels of poor water quality in 2010. This relationship holds across all models. Several possibilities may explain this finding. First, stakeholders in watersheds with a higher percentage of white residents may experience easier access to economic resources, greater political influence, and higher social capital. These advantages may contribute into better or newer infrastructure, for example, sewage, stormwater, more robust pollution prevention or control efforts, and more public programs or capacity that can improve water quality. A second possibility may be linked to environmental injustice conditions. Here, lower-income communities and/or those watersheds with a higher percentage of non-white individuals may be located closer to sites of contamination and/or such areas are less likely to be the recipients of spending designed to ameliorate those conditions and/or contaminants (Wang et al., 2024; Schaider et al., 2019)

4.5.4 A Word of Caution

Water quality changes are difficult to achieve, even those that report statistical significance. Shortly after sending an initial contact email to one survey participant in rural central Oregon working for a coalition of municipal water providers, she telephoned to talk about water quality.[2] In discussing water quality, she said that measurement is often difficult, indicating impacts from climatic patterns, changes in plumbing fixtures, and changes in household sizes. On the latter point, she noted that water quality is impacted when college students return home for the summer months as well as when they return to school. Even small factors like these make any measurement difficult. The goal for many working in watersheds – beyond litter clean-up programs and education – is not necessarily water quality improvement, according to this official, but rather to not let water quality worsen or, in other words, maintenance of present quality.

Another survey invitation provoked another telephone call; this time from a director of a local level coastal restoration nonprofit organization.[3] He indicated that larger organizations – those with more contacts and greater resources – are typically able to secure state funding for projects in various watersheds, often to the disadvantage of smaller, more scientifically focused organizations like his. This discussion provided a cautionary tale that state and regional level nongovernmental organizations can connect other organizations to produce positive environmental impacts. He indicated, however, that smaller organizations like his working in a single watershed on a specific task do not always benefit in terms of resources or a seat at the decision-making table when these larger, more powerful organizations are involved.

[2] Informal interview with official representative of a sampled organization, March 9, 2015.
[3] Informal interview with official representative of a sampled organization, April 17, 2015.

4.6 Conclusion

This chapter explores the relationships among revealed water quality outcomes and an array of variables that approximated external, civic and ecological conditions, and institutional/collaborative structures. Results paint a complicated picture; however, three key results are evident. First, regional organizations are important institutional actors associated with enhanced water quality outcomes. In fact, a one-unit increase in the number of regional organizations is associated with a 5% reduction in impaired waters. Second, historical water quality conditions matter a great deal and point to significant hurdles, should collaborations be seeking water quality restorations rather than watershed protection. In other words, poor water quality in 2002 is associated with poorer outcomes in 2010. Third, the percentage of individuals identifying as white in a watershed is associated with better water quality outcomes. The latter findings highlight the importance of understanding historical water quality baselines.

These exploratory results generate more questions than answers. Taking into consideration the number of individual organizations (or the composition of those organizations) that should be involved in multiple adjacent watersheds, should we begin to reconceptualize the watershed-by-watershed approach? If those working locally at the watershed level have little impact on actual water quality, even if they perceive they are having an impact, should watershed-related polices incentivize regional approaches? The evidence presented here is not sufficient to support this claim, but it does raise the question. On the contrary, perhaps incentives and nudges should focus on encouraging local organizations to connect with regional organizations so that knowledge, best practices, and other information can be shared across broad networks of interested organizations, connecting local actors together across watersheds and expanding information networks. Approaching the understanding of what occurs within and across watersheds allows us to examine further and offer suggestions about the approaches and techniques currently used to manage US water resources.

References

Berardo, Ramiro, and Scholz, John T. 2010. Self-Organizing Policy Networks: Risk, Partner Selection, and Cooperation in Estuaries. *American Journal of Political Science*, **54**(3), 632–649.

Berry, William, Fording, Richard, and Hanson, Russell. 2000. An Annual Cost of Living Index for the American States, 1960–1995. *Journal of Politics*, **62**(May), 550–567.

Brown, Mitchell, and Hale, Kathleen. 2014. *Applied Research Methods in Public & Nonprofit Organizations*. John Wiley & Sons.

Dillman, Don A., Smyth, Jolene D., and Christian, Leah Melani. 2014. *Internet, Phone, Mail, and Mixed-Mode Surveys: The Tailored Design Method*. Hoboken, N.J.: John Wiley & Sons.

Genskow, Kenneth, and Prokopy, Linda Stalker. 2009. Lessons Learned in Developing Social Indicators for Regional Water Quality Management. *Society & Natural Resources*, **23**(1), 83–91.

Hoornbeek, John. 2011. *Water Pollution Policies and the American States: Runaway Bureaucracies or Congressional Control?* Albany, NY: State University of New York Press.

Hoornbeek, John A. 2012. *Water Pollution Policies and the American States: Runaway Bureaucracies or Congressional Control?* State University of New York Press.

Imperial, Mark. 2005. Using Collaboration as a Governance Strategy: Six Lessons from Watershed Management Programs. *Administration and Society*, **37**(3), 281–320.

King, Gary. 1989. *Unifying Political Methodology: The Likelihood Theory of Statistical Inference*. New York: Cambridge University Press.

Kneese, Allen V. 2013. *The Economics of Regional Water Quality Management*. RFF Press.

Koontz, Thomas M., and Thomas, Craig. 2006. What Do We Know and Need to Know about the Environmental Outcomes of Collaborative Management? *Public Administration Review*, **66**, 111–121.

Long, J. Scott. 1997. *Regression Models for Categorical and Limited Dependent Variables*. Thousand Oaks, Calif.: Sage Publications.

Lubell, Mark. 2005. Do Watershed Partnerships Enhance Beliefs Conducive to Collective Action? In *Swimming Upstream: Collaborative Approaches to Watershed Management*, edited by Paul A. Sabatier, Will Focht, Mark Lubell, Zev Trachtenberg, Arnold Vedlitz and Marty Matlock, 201–232. Cambridge, MA: The MIT Press.

Luke, Douglas A. 2004. *Multilevel Modeling*. Thousand Oaks, Calif.: Sage Publications.

Morris, John C. 2022. *Clean Water Policy and State Choice: Promise and Performance in the Water Quality Act*. New York: Cambridge University Press.

Niles, Mereidth T., and Lubell, Mark. 2012. Integrative Frontiers in Environmental Policy Theory and Research. *Policy Studies Journal*, **40**(1), 41–64.

Ostrom, Elinor. 2005. *Understanding Institutional Diversity*. Princeton, NJ: Princeton University Press.

Ostrom, Vincent, and Ostrom, Elinor. 1972. Legal and Political Conditions of Water Resource Development. *Land Economics*, **48**(1), 1–14.

Poff, N. LeRoy, Allan, David, Bain, Mark B., Karr, James R., Prestegaard, Karen L., Richter, Brian D., Sparks, Richard E., and Stromberg, Julie C. 1997. The Natural Flow Regime: A Paradigm for River Conservation and Restoration. *BioScience*, **47**(11), 769–784.

Sabatier, Paul A., Focht, Will, Lubell, Mark, Trachtenberg, Zev, Vedlitz, Arnold, and Matlock, Marty (eds). 2005. *Swimming Upstream: Collaborative Approaches to Watershed Management*. Cambridge, MA: The MIT Press.

Schaider, Laurel A., Swetschinski, Lucien, Campbell, Christopher, and Rudel, Ruthann A. 2019. Environmental justice and drinking water quality: Are there socioeconomic disparities in nitrate levels in US drinking water? *Environmental Health*, **18**, 1–15.

Scheberle, Denise. 2004. *Federalism and Environmental Policy: Trust and the Politics of Implementation*. 2nd edn. Washington, D.C.: Georgetown University Press.

Schlager, Edella, and Blomquist, William. 2008. *Embracing Watershed Politics*. Boulder, CO: University Press of Colorado.

Scholz, John T., and Stiftel, Bruce (eds.). 2005. *Adaptive Governance and Water Conflict: New Institutions for Collaborative Planning*. Washington, D.C.: RFF Press.

Scholz, John T., Berardo, Ramiro, and Kile, Brad. 2008. Do Networks Solve Collective Action Problems? Credibility, Search, and Collaboration. *The Journal of Politics*, **70**(02), 393–406.

Singleton, Royce A., and Straits, Bruce C. 2010. *Approaches to Social Research.* 5th edn. New York: Oxford University Press.

Tijerina, Daniel. 2019. *HUC8 - CONUS Shapefile.* www.hydroshare.org/resource/b832a6c2f96541808444ec9562c5247e. Accessed 2025-07-31.

USGS. 2012. *Water Resources of the United States: Hydrologic Unit Maps.*

Wang, Runzi, Liu, Yiyi, Guan, Jianxing, Zuo, Chen, Dai, Congyi, Liu, Xiaofeng, Liang, Zhongyao, and Zhao, Gang. 2024. Environmental justice of Texas recreational water quality–The disproportionate E. coli levels and trends. *Journal of Environmental Management*, **370**, 122969.

Weber, Edward P. 2003. *Bringing Society Back in Grassroots Ecosystem Management, Accountability, and Sustainable Communities.* Cambridge, MA: The MIT Press.

Weible, Christopher M. 2010. Collaborative Institutions, Functional Areas, and Beliefs. In *Self-Organizing Federalism: Collaborative Mechanisms to Mitigate Institutional Collective Action Dilemmas*, edited by Richard C. Feiock and John T. Scholz, 179–203. New York: Cambridge University Press.

5

Findings Revealed: Actions and Outputs in Watershed Partnerships

Chapter 4 examined the associations between contextual, institutional, and collaborative variables and revealed water quality outcomes across 564 watersheds. Certain patterns emerged, particularly the significance of regional NGOs and the challenges related to historical uses and land use patterns. The analysis also underscored the limits of simple stakeholder counts as predictors of environmental outcomes. The presence of organizations alone does not ensure performance; it matters how those organizations are structured, how they make decisions, and what they do on the ground. This chapter investigates those internal dimensions of collaboration by examining the institutional design and operational activities of watershed partnerships.

To do so, we draw on a subsample of 167 watersheds for which detailed survey data were collected regarding collaborative practices. These data capture the specific actions partnerships undertake, the institutional arrangements they adopt, and the governance mechanisms they employ. Organized using Ostrom's (2005) Institutional Analysis and Development (IAD) framework, the analysis identifies collaborative activities at three distinct levels: constitutional-choice, collective-choice, and operational. By examining activities across these levels, we gain a clearer picture of how watershed collaborations attempt to build capacity, create and maintain legitimacy, share information, coordinate across actors and institutions, and pursue improved water quality via specific operational activities.

This chapter proceeds in four parts. First, we introduce the IAD framework and describe how collaborative activities were measured across the three institutional levels. Second, we present descriptive statistics for each category, highlighting patterns of institutional development, resource sharing, and on-the-ground implementation. Third, we assess the relationship between collaboration intensity and watershed performance using bivariate correlations and regression models. Finally, we discuss the implications of these findings for understanding the mechanisms through which collaboration shapes environmental outcomes.

In contrast to the broader multilevel models presented in Chapter 4, the analyses here are not intended to offer generalizable estimates across the full national sample. Rather, they provide a focused examination of how collaboration works in practice – what kinds of structures and behaviors are most common, how information flows and is shared, and whether more intensive forms of collaboration correspond with more legitimate and (ideally) better results.

5.1 Exploring Collaborative Activities: Institutional Levels and On-the-Ground Efforts

The policy output characteristics examined here operate at the action arena level of the IAD framework (Ostrom 2005), corresponding to the operational tier of the Levels of Collaborative Action Framework (LCAF) developed by Imperial (2005). Because Imperial identifies a wide range of concrete outputs at this level – many of which can be produced by diverse actors in a watershed setting – we disaggregate responses into two categories: environmental improvement activities and educative activities. These reflect distinct, yet complementary, contributions to watershed governance: The former directly shape ecological conditions, while the latter aim to build institutional capacity and public understanding. As shown in Table 5.1, each category encompasses multiple actions that respondents could report their organization as undertaking. These were collected via a simple checkbox format on the survey instrument.

To analyze the prevalence of these actions, we constructed composite indices following best practices for aggregating categorical variables into count-based

Table 5.1 *Levels of Collaborative Action Framework (LCAF) activities, adapted from Imperial (2005). Categories reflect key collaborative activities, examples, and their intended goals.*

Category	Examples of activities	Goals
Environmental Improvement	Habitat restoration Installation of best management practices (BMPs) Clean-up events	These activities directly target ecological outcomes and watershed quality.
Educative	Programs/events for schools Programs/events for elected officials Programs/events for public administrators, elected officials, other stakeholders	These activities aim to raise awareness and build capacity among key audiences and are more likely to indirectly impact water quality.

summary measures (Singleton and Straits 2010, 434–36). Each index is bounded between 0 and 1 and calculated using the formula:

$$\frac{\sum \bar{x}}{n},$$

where $\bar{x}$ represents a row vector of mean responses for each activity across all watersheds, and n is the total number of activities included in the index. The result captures the average density of environmental or educative engagement reported per watershed. These composite variables enter the models as outcome-relevant policy outputs, alongside the socioeconomic, civic, and ecological context variables described in Phase 1, as well as measures of state-level NPS activism and prevailing water rights doctrine.

While this analysis offers a more granular view of collaborative outputs, it is important to recognize the limitations imposed by the smaller sample size. These data provide valuable insights into how collaborations attempt to achieve their goals, but the findings cannot be assumed to represent the full national picture. For that reason, the multilevel models presented in the next section play a critical role in testing these relationships across the broader sample of 306 watersheds. Together, these analyses offer complementary perspectives: The collaborative activities analysis probes the internal workings of watershed partnerships and the institutional logic guiding their actions, while the multilevel models provide a more rigorous and generalizable test of whether the presence and scale of certain organizations, particularly those operating at regional levels, are associated with measurable environmental outcomes.

Taken together, these analyses reveal a critical finding: It is not the sheer number of collaborative activities undertaken that predicts improved water quality outcomes, but rather the presence and influence of organizations positioned to overcome scale-related challenges (Morris et al. 2013) and mobilize sufficient resources. In particular, the involvement of regional NGOs emerges as a consistently significant predictor of better outcomes, underscoring that collaboration's success hinges less on the quantity of activity and more on the institutional capacity and strategic positioning of key actors. This insight reframes how we assess collaborative governance, not as a simple matter of doing more, but as a question of whether the right organizations are involved and empowered to act effectively within complex, multi-scalar policy environments.

5.1.1 Descriptive Statistics of Collaborative Activities by IAD Level

Developed by Elinor Ostrom and colleagues in the 1990s, the IAD framework enables researchers and practitioners to examine how different types of decisions and decision-environments may shape collaborative and/or self-governing outcomes (Ostrom 1990; 1998, 1999, 2005, 2007). The focal point of the IAD framework is the action arena(s), which consist of two components: (1) a set of actors whose behavior is modeled after Simon's (1947) 'satisficing' man, and (2) a decision–action situation. For the former, key dimensions and assumptions relate to actors' goals and preferences, ability to process and organize information, decision-making heuristics, and available resources. For the latter, Sabatier (2005, 176) describes a context that provides and constrains who can participate, what moves and decisions are available, how much information is accessible, and the risks and rewards associated with participation. Within each action arena there is an action situation, which is the "social space where participants with diverse preferences interact, exchange goods and services, dominate one another, and fight..." (Ostrom 2005, 14). In this situation, the ability of stakeholders in a collaborative partnership to promote their policy goals is determined by the institutional rules in use, as well as by the material and biophysical world and the nature of the community involved.

In his analysis of six watershed management programs, Imperial (2005) adapted the IAD framework to create a LCAF. While the IAD framework primarily analyzes rules created at three levels – constitutional, collective, and operational – the LCAF acknowledges that collaborative activities occurring at each of these levels are interrelated (Imperial 2005, 288). From this perspective, there are three levels of joint action: operational, policy making, and institutional. These respectively correspond to aspects of the IAD framework: constitutional and collective choices, operational outputs, and shared perceptions and relationships.

Results are organized following the institutional delineations identified by the IAD framework: constitutional-choice, collective-choice, and operational, as defined in Ostrom's (2005) IAD framework. The next section discusses how many watershed partnerships engaged in each type of activity (out of the 167 watersheds in our sample), including the proportion of partnerships answering "Yes," "No," or "Not Sure" to whether they undertake that activity. In general, the data show that most watershed collaborations engage in multiple activities across all three levels, though specific types of activities vary widely in frequency.

5.1.1.1 Constitutional-Choice Level Activities

These high-level activities involve establishing the formal rules and plans guiding the collaboration. As Table 5.2 shows, an overwhelming majority of collaborations have developed some form of watershed plan, either by adopting an existing plan or producing a new one. Fully 139 watersheds (83) have a watershed management plan in place, and nearly 79 specifically reported that the collaboration itself produced a watershed plan. Likewise, nearly two-thirds (63) have formalized

Table 5.2 *Constitutional-choice level activities in watershed partnerships. Activity categories are adapted from Imperial (2005). Values represent the frequency and percentage of partnerships producing each type of activity.*

Activity (formal output)	Yes (count, %)	No (count, %)
Watershed plan (any in place)	139 (83%)	28 (17%)
Produced a watershed plan	132 (79%)	35 (21%)
Memorandum of understanding (MOU)	105 (63%)	62 (37%)
Capital improvement plan	95 (57%)	72 (43%)
Land use plan	86 (51%)	81 (49%)
Harbor plan	7 (4%)	160 (96%)
Conflict-resolution mechanisms	48 (29%)	119 (71%)
Legal agreements (e.g., contracts)	74 (44%)	93 (56%)
Influenced state/federal legislation	68 (41%)	99 (59%)

their partnership through a memorandum of understanding (MOU) or similar agreement. Other constitutional-level outputs are somewhat less common: For example, about 41% of collaborations produced or influenced state/federal legislation related to water management, and 29% established formal conflict-resolution mechanisms within their institutional design. Only a small subset of cases pursued very specialized outputs – for example, only 7% of collaborations (~4) developed a harbor or port plan (unsurprising given not all watersheds include navigable harbors). In a few instances, respondents selected "Not Sure," indicating uncertainty about certain high-level outputs (e.g., whether some "other" unspecified constitutional output was produced). In general, however, most respondents were aware of their collaboration's formal arrangements, so "Not Sure" responses were minimal at this level (treated as a negative response in our analysis). On average, each watershed collaboration undertook about 4 to 5 distinct constitutional-level activities (median = 4, range = 0–10), indicating a moderate degree of formal institutional development in most partnerships.

5.1.1.2 Collective-Choice Level Activities

These activities relate to how stakeholders make decisions together, share resources, and coordinate their efforts. Our survey included a wide array of possible collective-choice activities, reflecting the many ways partners can collaborate beyond formal rulemaking. Table 5.3 presents the frequencies for these activities which we group into two subcategories for clarity: (a) Joint Planning and Policy Consultations (formal coordination on goals, standards, budgets, etc.) and (b) Resource and Information Sharing activities (sharing staff, funds, data, and knowledge across organizations).

Table 5.3 *Collective-choice level activities in watershed partnerships. Activity categories are adapted from Imperial (2005). Values represent the percentage of partnerships reporting each type of activity.*

Consultation activity	Yes (%)	No (%)
Habitat restoration priorities	87%	13%
Infrastructure investment plans	77%	23%
Land acquisition priorities	55%	45%
Formal shared goals	72%	28%
Formal shared policies	47%	53%
Joint budgeting / financing	66%	34%
Formal monitoring protocols	68%	32%
Water quality standards (formal)	61%	39%
Water quantity standards (formal)	63%	37%
Environmental enhancement goals	65%	35%
Progress reports on goals	64%	36%

Note: Percentages may not sum to 100 due to rounding. "No" indicates the partnership did not report activity in that category.

A large proportion of watershed partnerships engage in collective decision-making on substantive issues and joint policies. As shown in Table 5.3, 87% of collaborations consult on habitat restoration priorities, making it the most common collective-choice activity at this level. Nearly as many (around 72%) have discussed establishing formal shared goals for the watershed, and over 46% have worked toward formal shared policies or regulations. Other frequent planning-level collaborations include consulting on monitoring protocols (≈68%), joint budgeting/finance (≈66%), and setting water quality or quantity standards (each around 60–65% of partnerships). In general, most partnerships undertake multiple such consultations, often aligning on goals, plans, and science as part of their collaborative process. For example, about 54% discuss land acquisition priorities, and 19% report some "other" consultation activity not listed (e.g., perhaps consulting on recreation or flood control plans). Notably, "Not Sure" responses were somewhat more common here (though still relatively low); for instance, a few respondents were unsure if their partnership had, say, a formal joint policy or not. These uncertain cases (generally on the order of 5–10 or less) are again treated as no/absent in counts. On average, each watershed group engaged in 8–9 of the 12 listed joint planning activities, indicating a high level of collective decision-making across the sample.

The second subcategory of collective-choice activities focuses on how partnerships share resources and knowledge to enhance their collaborative capacity. As shown in Table 5.4, these activities are also quite prevalent. For example,

Table 5.4 *Additional collective-choice level activities in watershed partnerships. Values represent the percentage of partnerships reporting each type of activity.*

Collective-choice activity	Yes (%)	No (%)
Joint research and fact-finding	83%	17%
Shared database or data sharing	76%	24%
Shared technical information	85%	15%
Shared technical resources (equipment, etc.)	80%	20%
Co-locating staff across organizations	67%	33%
Collecting information for other organizations	78%	22%
Ad-hoc knowledge sharing groups	82%	18%
Informal information sharing (networking)	80%	20%
Formal collaborative information exchange	74%	26%
Shared current events/news	90%	10%
Shared opportunity knowledge (e.g., grants)	82%	18%
Hired staff for another organization	21%	79%
Shared volunteers with other organizations	58%	42%
Pooling financial resources (joint projects)	76%	24%
Funding other organizations' priorities	59%	41%
Providing staff support to other agencies	67%	33%

Note: Percentages may not total 100% due to rounding. "No" indicates the partnership reported no participation in that category. Activity categories are adapted from Imperial (2005).

85% of collaborations share technical information or expertise, and 80% share technical resources such as equipment or facilities. Joint research or fact-finding initiatives are conducted by 83% of partnerships, indicating a strong emphasis on building shared knowledge bases. Informal information-sharing groups (e.g., ad-hoc committees or networks) are utilized by 82%, while 74% participate in formal collaborative knowledge-sharing platforms (e.g., coalitions or committees). We see a high incidence of knowledge sharing mechanisms. Notably, 90% share current events or news relevant to the watershed among partners, and 82% coordinate through ad-hoc working groups at times. Formal knowledge networks are also common: about 74% have participated in a formal collaborative knowledge-sharing platform (e.g., a formal coalition or committee structure), and 75% have some form of shared data systems or databases linking the partners. In addition, many collaborations share tangible resources: About 58% report sharing volunteers across organizations, 59% provide funding to support other organizations' projects or priorities, and 76% pool financial resources for joint activities. Around 67% have even shared or supported staff across organizations (e.g., assigning staff from one entity to assist another), and 21% went as far as hiring staff to work in another organization (this last being less common due to the complexity of such arrangements).

Overall, the resource-sharing activities show that most watershed collaborations coordinate closely, providing staff support, pooling money, and exchanging technical resources (about 85% share technical information or expertise). "Not Sure" responses were slightly more common on these items (some respondents were unsure, for instance, if their partners had ever co-located staff or not), but in general the patterns are clear: Nearly all collaborations engage in extensive information exchange, and many also share personnel or funds across organizational boundaries. On average, each watershed partnership performed 16 out of 30 collective-choice activities measured, with a median of 17, underscoring that collaborative watershed governance typically involves a dense network of resource-pooling and knowledge-sharing among participants.

5.1.1.3 Operational Level Activities

We next grouped operational activities into two broad types: education or outreach activities and restoration or management interventions. Table 5.5 displays the frequencies. The data indicate that virtually every watershed group engaged in at least some operational projects. In terms of education and outreach, about 87% of collaborations hosted general outreach programs or events for the public, and over 81% organized special events like public workshops or conferences to raise awareness. Many also targeted specific audiences: roughly 75% ran school-based education programs for youth, 74% provided homeowner education (e.g., on septic maintenance or landscaping to protect water), and 42% even did outreach to industry or agricultural producers about best practices. Training programs for stakeholders were conducted by 65% of partnerships. Overall, public education is a core focus of most watershed collaborations. The median number of distinct outreach activities per watershed is 8 out of 11 types surveyed.

On the restoration and management side, we see similarly widespread action: 91% of collaborations implemented habitat restoration projects (such as streambank restoration or wetland enhancement), and 86% organized stream or shoreline cleanup activities. A majority also tackled nonpoint source pollution through best management practices (BMPs): 61% worked on agricultural BMP installations, 52% on urban stormwater BMPs, and 48% on forest management BMPs, reflecting the diverse land-use contexts of different watersheds. Over half (55%) pursued land acquisition or conservation easements to protect water resources. Fewer groups engaged in heavy infrastructure projects: only ~6 reported installing new sewers or wastewater infrastructure, and about 10 helped facilitate upgrades of septic systems (OSDSs), likely because such projects fall to utilities or require more funding than most collaborations have.

Table 5.5 *Participation rates in education, outreach, and restoration activities by watershed partnerships. Activity categories are adapted from Imperial (2005). Values represent the percentage of partnerships engaging in each activity.*

Operational activity	Yes (%)	No (%)
Education and outreach		
Outreach programs/events (general public)	87%	13%
Special public events/conferences	82%	18%
School-based education programs	75%	25%
Homeowner education programs	74%	26%
Industry education programs	42%	58%
Training programs (capacity building)	65%	35%
Education for elected officials	48%	52%
Education for public administrators	51%	49%
Public social events (community-building)	74%	26%
School curriculum development	45%	55%
Other education/outreach activity	16%	84%
Restoration and management		
Habitat restoration projects	91%	9%
Clean-up activities (trash or debris removal)	86%	14%
Land acquisition/conservation	55%	45%
Installing urban BMPs (e.g., rain gardens)	52%	48%
Installing agricultural BMPs	61%	39%
Installing forest BMPs	48%	52%
Installing water flow BMPs (e.g., dams)	48%	52%
Installing/upgrading sewers	6%	94%
Upgrading septic systems (OSDS)	10%	90%

Note: Percentages may not total 100% due to rounding. "No" indicates the partnership reported no activity in that category.

Overall, the typical watershed partnership performed approximately 12 operational actions out of 22 surveyed (median = 12, range 0–22), with each collaboration focusing on a unique mix of outreach and restoration. "Not Sure" responses at this level were rare (respondents usually knew if a project was done or not), so the table reflects clear yes/no tallies. These results highlight that operational outputs are the most immediately visible aspect of collaboration, and nearly all partnerships undertake multiple projects aimed at improving water conditions in their watershed.

In summary, our descriptive results illustrate that watershed collaborations are multifaceted: most engage in a mix of high-level institutional arrangements (plans and agreements), collaborative governance mechanisms (shared staffing, joint decision-making, and knowledge exchange), and locally implemented projects (education campaigns, restorations, and BMP installations). The typical partnership in our sample undertook approximately 4–5 constitutional-choice activities, about 16 collective-choice activities, and around 12 operational activities. We now turn to whether these aggregate measures of collaboration are associated with environmental outcomes – specifically, whether greater collaboration at each level corresponds with improved water quality.

5.1.2 Collaboration Indices and Bivariate Correlations with Water Quality

To facilitate analysis, we created a simple count index for each IAD level representing the total number of activities a collaboration conducted. For example, if a partnership answered "Yes" to k out of n constitutional-level items, its Constitutional Collaboration Index would be k. Similarly, we counted the number of collective-choice activities (out of $n_{cc} = 30$) and operational activities (out of $n_{op} = 22$) for each watershed. These indices provide an overall gauge of collaboration intensity at each level. In our sample, the Constitutional Index ranged from 0 to 10 (mean $\approx$4.4), the Collective Index ranged from 0 to 29 (mean $\approx$16.2), and the Operational Index ranged from 0 to 22 (mean $\approx$11.6). This confirms that there is substantial variation in how extensively different collaborations operate at each level; some are relatively minimalist, while others are very active on many fronts.

Before controlling for other factors, we examined the bivariate relationship between these collaboration indices and the watershed's water quality outcomes. Our outcome measure is the Percent of Insufficient Waterbodies in 2010, denoted as Y_{2010}, which represents the proportion of assessed water bodies in the watershed classified as impaired (not meeting water quality standards) in the year 2010. Higher values indicate worse water quality. If collaboration is yielding environmental benefits, we might expect a negative correlation (i.e., watersheds with more collaborative activities have a lower percent of impaired waters). Table 5.6 reports the Pearson correlation coefficients r between each index and the 2010 outcome.

As shown in Table 5.6, none of the collaboration indices exhibits a statistically significant linear correlation with water quality outcomes. The Pearson correlation

Table 5.6 *Bivariate correlations of collaboration activity indices with the percentage of impaired waters in each watershed (2010).*

Collaboration activity index	Correlation with % impaired waters (2010)
Constitutional-level activities (count)	+0.03 (n.s.)
Collective-choice activities (count)	−0.02 (n.s.)
Operational activities (count)	−0.05 (n.s.)

Note: "n.s." indicates correlation is not statistically significant at conventional thresholds.

coefficients are all very close to 0 and not statistically different from 0. Specifically, the Constitutional Index has a slight positive correlation with the percent of impaired waters ($r \approx +0.03$, $p > 0.1$), whereas the Collective and Operational indices show very small negative correlations ($r \approx -0.02$ and $r \approx -0.05$, respectively; both $p > 0.1$). In practical terms, this means that watersheds with higher values on the collaboration indices did not, on average, have noticeably better water quality outcomes in 2010 than those with lower values. For example, a watershed partnership that implemented many projects and plans (i.e., with a high index value) could still have a similar percentage of impaired waters as a partnership that was less active. This lack of bivariate association suggests that any benefits of collaboration may be subtle or confounded by other factors; indeed, it is possible that highly impaired watersheds spur more collaboration, which could produce a positive raw correlation between collaboration count and impairment (or cancel out any negative correlation from improvement). The bivariate results therefore underscore the importance of controlling for underlying differences between watersheds.

5.1.3 Regression Analysis: Collaboration Indices and Water Quality Outcomes

We next employed regression models to assess whether collaboration at each level is related to water quality outcomes when accounting for key contextual variables. Specifically, we estimated ordinary least squares (OLS) and multilevel models predicting the 2010 percent of insufficient (impaired) water bodies in each watershed. All models include controls for the watershed's prior water quality in 2002 (a lagged outcome measure), demographic composition, socioeconomic factors, and population density, among others. By including the 2002 impairment percentage, we effectively control for baseline problem severity – a crucial factor, since watersheds with more serious pollution problems may both have worse 2010 outcomes and mount more collaborative efforts. The demographic controls capture potential socio-economic drivers of water quality (e.g., development patterns, resources, or legacy pollution): Percent White population (2000), Socioeconomic Status (SES) measured as the proportion of high-income households, and Population Density (2000). These variables help isolate the impact of collaborative activities from structural advantages or disadvantages a watershed might have.

Table 5.7 presents the results of two representative regression models: Model 1 is a standard OLS regression, and Model 2 is a two-level random-intercept model (watersheds nested within states). The multilevel model accounts for the possibility that watersheds in the same state share unobserved characteristics (e.g., state policies or climate). The state-level variance component was significant in our data, so we report the multilevel estimates as our preferred specification. For

Table 5.7 *Regression estimates for the association between collaboration indices and water quality outcomes. Model 1 is OLS; Model 2 is a two-level multilevel model with watersheds nested in states.*

Predictor	Model 1: OLS coefficient (SE)	Model 2: Multilevel coefficient (SE)
Collaboration activity indices		
Constitutional-level activities (count)	+0.002 (0.005)	+0.001 (0.005)
Collective-choice activities (count)	−0.001 (0.002)	−0.001 (0.002)
Operational activities (count)	−0.003 (0.003)	−0.004 (0.003)
Baseline and control variables		
Percent impaired (2002)	+0.600*** (0.060)	+0.587*** (0.055)
Percent white population (2000)	+0.180** (0.065)	+0.172** (0.060)
Socioeconomic status (SES, 2000)	−0.220* (0.105)	−0.213* (0.099)
Population density (ln, 2000)	−0.015* (0.007)	−0.014* (0.006)

Note: Standard errors in parentheses. Significance levels: $^{*}p < 0.05$, $^{**}p < 0.01$, $^{***}p < 0.001$.

brevity, we show the coefficients for the collaboration indices and key controls; coefficients for additional control variables (e.g., watershed size, presence of interstate compact) are omitted here as they were not central to the hypothesis. Both models yield very similar findings, so our discussion focuses on the substantive results common to both.

Two clear patterns emerge from Table 5.7. First, the baseline water quality and sociodemographic controls explain most of the variance in 2010 outcomes, while the collaboration indices have no statistically significant effects. The coefficient on Percent Impaired in 2002 is approximately +0.59 ($p < 0.001$), meaning that, holding other factors constant, a 10 percentage-point higher impairment rate in 2002 is associated with roughly a 6 percentage-point higher impairment in 2010. This underscores the inertia in water quality conditions: watersheds that started off in poor condition tended to remain in relatively poor condition in 2010, even after a decade which included collaborative interventions and activities, although such activities may have prevented further watershed deterioration. Among the static characteristics, Percent White population is positively associated with impairment in 2010 ($\beta \approx +0.17$, $p < 0.01$ in Model 2). Substantively, watersheds with more predominantly white populations had better water quality outcomes, a pattern consistent with historical industrial land use or environmental inequities (i.e., these watersheds may have legacy pollution issues). Higher socioeconomic status correlates with better water quality ($\beta \approx -0.21$, $p < 0.05$): wealthier communities are likely to enjoy greater resources for pollution control or less polluting land uses, leading to lower impairment percentages. Population density also shows a mild beneficial effect ($\beta \approx -0.014$, $p < 0.05$), suggesting that more urbanized watersheds tend to have slightly lower fractions of impaired water bodies once we account for other factors. This finding aligns with the idea that urban areas often invest in better waste treatment and infrastructure, mitigating certain pollution problems – an interpretation echoed by the positive bivariate correlations observed between urbanization and water quality improvement in our data.

Second, and central to our research question, none of the collaboration activity indices show a significant impact on the 2010 water quality outcomes once controls are included. The coefficients for the Constitutional, Collective, and Operational indices are all very small and statistically non-significant. In Model 2 (our preferred multilevel model), the point estimates suggest that an increase of one additional activity at any level is associated with at most a 0.1–0.4 percentage-point change in impaired waters (and even those tiny effects could be in either direction: The Constitutional Index has a positive sign, while Collective and Operational are negative, but again none differs from 0 in a meaningful way). In practical terms, a collaboration's sheer number of plans, agreements, committees, or projects did not independently predict better water quality. This result holds in OLS and in the state-clustered model, indicating it is robust to accounting for state-level heterogeneity. We also tested variations of these models (not shown in the table) such as including the indices one at a time or using alternative functional forms (e.g., a combined index of all activities, or categorical "high vs. low collaboration" indicators); none yielded a significant relationship between collaboration intensity and outcomes.

It is important to emphasize that these null findings do not imply that collaborative efforts lack value. Rather, they suggest that any positive impacts of collaboration on water quality are either too small to detect at this broad scale or are being masked by the influence of prior conditions and external factors. Moreover, our findings do not measure waters protected, social capital generated, scientific expertise shared, political efficacy gained, or new organizational capacities, all of which may lead to future environmental benefits. These findings are especially important given that collaborations often operate in watersheds facing the greatest challenges; indeed, many partnerships form precisely because of severe water quality problems. Our controlled regression confirms that initial impairment is a dominant predictor of later impairment, which can overshadow incremental improvements, waters protected, or even where water was kept from further degradation. In fact, as some respondents highlighted, many collaborative activities (especially restoration projects or behavior change programs) may take years or decades to translate into measurable water quality gains. For instance, installing riparian buffers or educating the public about fertilizer use might gradually reduce nutrient runoff, but the effect might not register strongly by the 2010 assessment.

Another consideration is that projects may affect relatively small segments of a waterway. It may be that water quality in that specific area improves, but the effect in the larger measurement area in the waterway is too small to be captured in larger-scale water quality measurements. An example of this is reported in Morris et al. (2013), in which stakeholders collaborated to execute the Elizabeth River Project, which sought to carry out a project to restore a specific location on the Elizabeth River, known as Money Point. Money Point was the site of a former creosote plant that had spilled large quantities of creosote into the river, creating a segment (about 150 meters) of very toxic river bottom. The project removed the contaminated river bottom, which immediately improved water quality at that point. However, the project had very little effect on the 25-mile length of the river. Additionally, several collaboration activities focus on preventing further decline rather than outright improvement. An official from Oregon, for example, noted in our survey that maintaining current water quality (preventing it from getting worse) is often a key goal, even if water quality does not dramatically improve. Such preventative successes would not show up as a decrease in percent impaired (the metric would remain stable), but they are still valuable outcomes.

Another perspective is that not all collaboration is equal. The type and scale of collaborative engagement may matter more than the quantity. Our indices treated each activity equally, but it could be that certain activities (or the involvement of certain actors) are the real game-changers. For example, we found earlier that involvement of regional organizations was correlated with better outcomes. This implies that collaboration spanning larger geographic scales or bringing in higher-level resources might yield more impact than collaboration confined to local stakeholders. In our broad indices, those nuances disappear; a watershed could score high on collective activities by sharing lots of information among local groups, but if it lacks connections to state/regional actors or the capacity to implement changes, the water outcomes may remain unchanged. Indeed, one interviewee cautioned that smaller local groups may struggle to access resources or influence policy without larger organizations at the table, even though those larger organizations can sometimes overshadow local voices. This hints that the *quality* of collaboration (e.g., bridging local and regional scales) is a crucial factor.

The regression analysis indicates that when accounting for initial conditions, demographics, and other factors, we do not find evidence that simply doing "more" collaborative activities at any level guarantees better water quality outcomes by 2010. Collaborations with extensive activities performed no better, on average, than those with fewer activities, once we factor in how impaired the watershed was to begin with. This is a notable finding: It suggests that the relationship between collaboration and environmental performance is not a straightforward linear one where more collaboration automatically equals cleaner water. Instead, it appears to be complex and conditional, potentially depending on *what kinds* of collaborative actions are taken and *who* is involved, as well as on the timescale of outcome realization.

5.2 Collaboration Type, Scale, and Environmental Performance

The previous analyses offer useful insights into how collaborative activity relates to watershed outcomes, but they also raise critical questions about the limits of such efforts. Most watershed partnerships in our sample engage in a diverse array of governance activities, drafting watershed plans, coordinating across agencies, and implementing on-the-ground interventions. Yet these efforts, in the aggregate, were not associated with measurable reductions in the percentage of impaired water bodies over the study period. This finding should not be read as evidence that collaboration is ineffective. Rather, it reflects the extraordinary difficulty of achieving environmental improvement, particularly in places defined by pollution legacies, fragmented land use, and limited institutional capacity. Many collaborations, it seems, are focused less on immediate reversal of degradation and more on containment, slowing decline, preventing further harm, and preserving existing gains. In that context, the absence of observable improvement may itself represent a form of success, albeit one not captured by conventional performance metrics.

At the same time, the lack of significant relationships between collaboration volume and outcomes suggests that more activity is not inherently better. A partnership may undertake dozens of initiatives but still fall short if it lacks the authority, alignment, or resources to target dominant pollution sources or larger segments of a body of water. Conversely, a collaboration with fewer but more strategically chosen actions, such as securing conservation easements or driving policy reform, may achieve greater impact. What matters, in other words, is not quantity but focus: Effective collaboration hinges on choosing the right activities, at the right scale, and with the right partners in place.

Among the few consistent signals in the data is the role of regional organizations. Their involvement correlates with better outcomes, pointing to the importance of institutional scale and cross-jurisdictional capacity. While local actors build trust and mobilize community engagement, they often lack the reach or resources to address systemic issues. Regional NGOs and intergovernmental entities, by contrast, can coordinate across boundaries, broker resources, and align governance efforts with the scope of ecological problems. This is consistent with polycentric governance theory and underscores the importance of vertical integration, connecting local initiative with regional and state-level infrastructure and support.

In short, collaboration appears to be a necessary but insufficient condition for improving water quality. It can create the organizational scaffolding needed for progress (e.g., shared goals, mutual trust, and coordinated action) but it cannot overcome structural and ecological constraints on its own. Without investments in infrastructure, policy enforcement, and land use reform, even the most well-intentioned partnerships may struggle to deliver measurable change. The policy lesson is clear: Collaborative governance should be seen not as a substitute for substantive environmental policy, but as a complement to it.

Finally, these findings invite future research to move beyond surface-level metrics – counts of meetings, plans, or participants – and instead examine the specific institutional pathways through which collaboration translates into outcomes. Understanding *how* collaboration works, under what conditions, and through which mechanisms is the next frontier. That task begins in Chapter 6, where we return to the full national sample to explore whether specific organizational configurations, especially those involving higher-capacity regional actors, are systematically associated with improved watershed performance.

5.3 Conclusion

This chapter has shown that collaboration, while widespread and multi-layered, does not automatically produce better water quality outcomes. The intensity of collaborative activity, measured across constitutional, collective, and operational dimensions, bears little statistical relationship to environmental performance once prior conditions are accounted for. What matters more, we find, is who is involved and how they are positioned within the institutional landscape. In particular, the presence of regional organizations stands out as a potential lever for improving outcomes, offering the scale, capacity, and connectivity that many local efforts lack.

These findings complicate the narrative that collaboration matters. They suggest instead that collaboration must be strategically structured and aligned with ecological scale, supported by meaningful institutional resources, and integrated across levels of governance. In the chapters that follow, we explore this proposition in greater depth, asking which types of organizational configurations are most conducive to meaningful environmental change.

References

Imperial, Mark. 2005. Using Collaboration as a Governance Strategy: Six Lessons from Watershed Management Programs. *Administration and Society*, **37**(3), 281–320.

Morris, John C., Gibson, William A., Leavitt, William M., and Jones, Shana C. 2013. *The Case for Grassroots Collaboration: Social Capital and Ecosystem Restoration at the Local Level*. Lanham, MD: Lexington Books.

Ostrom, Elinor. 1990. *Governing the Commons: The Evolution of Institutions for Collective Action*. Cambridge, MA: Harvard University Press.

Ostrom, Elinor. 1998. A Behavioral Approach to the Rational Choice Theory of Collective Action. *American Political Science Review*, **92**(1), 1–22.

Ostrom, Elinor. 1999. Institutional Rational Choice: An Assessment of the Analysis and Development Framework. In: Sabatier, Paul A. (ed.), *Theories of the Policy Process*. Boulder, CO: Westview Press.

Ostrom, Elinor. 2005. *Understanding Institutional Diversity*. Princeton, NJ: Princeton University Press.

Ostrom, Elinor. 2007. A Diagnostic Approach for Going beyond Panaceas. *Proceedings of the National Academy of Sciences*, **104**, 15181–15187.

Sabatier, Paul A., Leach, William D., Lubell, Mark, and Pelkey, Niel W. 2005. Theoretical Frameworks Explaining Partnership Success. In *Swimming Upstream: Collaborative Approaches to Watershed Management*, edited by Paul A. Sabatier, Will Focht, Mark Lubell, Zev Trachtenberg, Arnold Vedlitz, and Marty Matlock, 173–199. Cambridge, MA: The MIT Press.

Simon, Herbert. 1947. *Administrative Behavior: A Study of Decision-Making Processes in Administrative Organization*. New York: The Macmillan Company.

6

Perceptions of Water Quality Improvement

The perceptions of participants in a collaborative environment are also important. Certainly, perceived water quality also shapes conversations relating to public health, environmental and ecological sustainability, and economic development. Here too, there are likely a variety of factors that influence how stakeholders and groups engaged in the difficult work of collaboration perceive their effectiveness and if they have reached, or are reaching, their goals. These perceptions, we should note, are also shaped by actors' interpretations of relevant state and federal laws, other actors and their priorities, and the historical experiences of those collaborating (or attempting to collaborate). While the Clean Water Act (CWA) has contributed to improved water quality throughout the United States, progress is uneven and increasingly challenged by a host of anthropogenic and naturally occurring factors.

In this context, and with these external factors, perceptions of water quality take on greater importance. Perceiving improvements in water quality, for example, can contribute to heightened levels of community support, participation, and momentum, as well as increased participation in watershed activities, education, and awareness, and levels of political efficacy. For elected officials, this sort of "momentum" can lead to greater support for water policies, resources, and/or use of water policy instruments. Improvements in water quality perceptions may also contribute to increased usage of the watershed, better compliance outcomes, and overall attitudes regarding the body of water and, in turn, greater collaboration in the future. Conversely, negative beliefs may undermine levels of political efficacy, diminish effectiveness and engagement among watershed users and elected officials, regardless of the revealed water quality outcomes. This chapter recognizes the importance of perceptions and focuses on perceived improvements, levels of legitimacy and trust among actors, and whether perceptions are, in fact, related to revealed outcomes.

Table 6.1 *Dependent variable used in the statistical analysis, based on respondent perceptions of change in water quality since partnership involvement.*

Variable	Type	Description
Dependent variable: Watershed outcomes	Dependent variable	Watershed outcome based on respondent perception
Perceived water quality improvement	Dichotomous*	Respondent's perception of water quality improvement since involvement (0 = No; 1 = Yes)

*Coded as a binary outcome variable.

At the outset, however, it is worth mentioning that the sample size used for this part of the analysis is considerably smaller than that examined in the previous chapter; this phase includes information from respondents in 114 watersheds, rather than the 306 that were included in the first phase. This smaller sample size is because complete responses to the survey questions used in this phase were considerably lower. Even still, the watersheds discussed in this section are in each of the 48 contiguous states as well as Alaska; there were no responses from organizations operating in watersheds in Hawaii. Stakeholder perceptions of water quality in their watershed is the focus of this section and, indeed, perceptions of water quality improvement form the dependent variable in the hypotheses and quantitative analysis presented and described in the following subsections.

6.1 Dependent Variable

The dependent variable for this chapter relates to perceived water quality improvement, as indicated in the model presented in Chapter 3. This measure is indicated in Table 6.1. Our survey asked respondents to think back over time since their organization's involvement in watershed-related activities and indicate – yes or no – if they *perceive water quality improvement*. Each of these perceived outcomes is measured as dichotomous variables with our covariates presented in Tables 6.2–6.5.

6.2 Independent Variables

As shown in Table 6.2, we asked survey respondents about ecological conditions. These questions provide information about whether all, most, some, or none of the river segments in the watershed are naturally flowing. For analysis purposes, we use a dichotomous indicator of *no natural flowing waters*, where 0 indicates the watershed has some level of natural flowing and 1 means wholly altered

Table 6.2 *Socioeconomic and ecological contextual variables for each watershed.*

Variable	Type	Description
Socioeconomic status	Continuous*	Percentage of households in the watershed with annual incomes of $75,000+
Percent white population	Continuous*	Percentage of white population in the watershed
Population density	Continuous	Persons per square mile
Watershed size	Continuous	Logged square miles
No natural flowing waters	Dichotomous	Indicates whether the watershed has some or all naturally flowing waters (0) or a completely altered system with no natural flow (1)
Natural flow regime	Dichotomous	Indicates if a flow regime is in place to mimic natural flow in riverine segments (0 = No; 1 = Yes)
Intermittent	Continuous*	Percent of riverine water with seasonal/intermittent flow
State water rights doctrine	Dichotomous	State water law (0 = Riparian Rights; 1 = Prior appropriation)
National Estuary Program (NEP)	Dichotomous	Indicates whether the watershed is part of the NEP (0 = No; 1 = Yes)

*Logged to adjust for skew.

(unnatural) flow. If all segments do not flow naturally, a dichotomous indicator asks if a *natural flow regime* is in place to mimic a natural flowing river.

Variables relating to the stakeholder community are listed in Table 6.3. These include a dichotomous indicator for the presence of a watershed partnership. To measure political efficacy, the survey asks respondents to indicate how significant a role their organization has in watershed management decisions, measured on a Likert scale ranging from very insignificant (1) to very significant (5).

Table 6.4 includes measures of trust, beliefs, and legitimacy. Respondents were asked to indicate their level of trust with other watershed stakeholders by providing a level of trust in government stakeholders and trust in nongovernment stakeholders, measured on a scale of 1 to 10, with 1 indicating complete distrust and 10 indicating complete trust.

Three indicators of collective action beliefs are also measured. First, the question for a problem diffusion variable, measured on a scale of 1–100%, asks respondents to indicate the percentage of all watershed stakeholders that would have to change behaviors to resolve water-related problems.[1] Second, two

[1] This problem diffusion measure is commonly used in collective action studies (see Lubell et al. 2002; 2005).

Table 6.3 *Stakeholder composition and institutional variables used in the statistical analysis.*

Variable	Type	Description
Nongovernmental stakeholders	Count	Number of nongovernmental stakeholders, separated by types as indicated in Table 1.2
Governmental stakeholders	Count	Number of governmental stakeholders, separated by types as indicated in Table 1.2
Watershed partnership	Dichotomous	Indicates whether a watershed partnership is present in the watershed (0 = No; 1 = Yes)
Political efficacy	Scalar*	Indicates how significant survey respondent organizations perceive their role in watershed management processes (1 = Very Insignificant; 5 = Very Significant)

*Survey-based Likert or scalar measure.

measures of scientific knowledge are included. Lubell (2005, 206) writes, "Because transaction costs are a function of uncertainty, they are high when stakeholders perceive scientific knowledge as inadequate." The survey asks respondents to indicate their level of scientific understanding about the nature of the resource (Natural Scientific Understanding) and the level of scientific understanding about the causes of watershed problems (Causal Scientific Understanding), both of which are measured on a scale of 1 to 10, with 1 indicating very inadequate and 10 indicating very adequate understanding.

Two survey questions capture legitimacy characteristics. The first measures beliefs about watershed management decision-making fairness, asking whether the process is fair to all stakeholders. The second measures process representativeness, asking whether the management process adequately represents the respondent's organizational interests and concerns. Both are measured on a Likert scale ranging from strongly disagree (1) to strongly agree (5).

Finally, Table 6.5 summarizes two policy output variables reported by organizations. These include a composite measure of improvement and restoration activities – such as habitat restoration, riparian clean-up, and best management practice (BMP) installation – and a measure of education programs and events targeting various audiences from schools to elected officials.

Together, these variables capture the institutional context, stakeholder composition, belief systems, and reported activities of watershed partnerships. Grouping them thematically clarifies how each dimension contributes to the broader analysis of collaborative performance.

Table 6.4 *Variables capturing stakeholder trust, collective action beliefs, and perceptions of legitimacy.*

Variable	Type	Description
Trust in governmental stakeholders	Scalar*	Level of trust survey respondents have in governmental watershed stakeholders (1 = Complete distrust; 10 = Complete trust)
Trust in nongovernmental stakeholders	Scalar*	Level of trust in nongovernmental stakeholders (1 = Complete distrust; 10 = Complete trust)
Problem diffusion	Scalar*	Percent of all stakeholders who would have to change behavior to resolve water-related problems
Natural scientific understanding	Scalar*	Perceived scientific understanding of natural processes (1 = Very inadequate; 10 = Very adequate)
Causal scientific understanding	Scalar*	Perceived understanding of causal relationships related to watershed problems (1 = Very inadequate; 10 = Very adequate)
Decision-making fairness	Ordinal*	Agreement that decision-making is fair to all stakeholders (1 = Strongly disagree; 5 = Strongly agree)
Process representativeness	Ordinal*	Agreement that the process represents respondent's organization's interests (1 = Strongly disagree; 5 = Strongly agree)
Political efficacy (Perceived)	Ordinal*	Perception that the organization has a significant role in decision-making (1 = Very insignificant; 5 = Very significant)

*Survey-based Likert or scalar measure.

Table 6.5 *Variables capturing policy outputs reported by watershed organizations.*

Variable	Type	Description
Improvement/Restoration activities	continuous	Composite measure of the proportion of organizations engaged in restoration, clean-up, BMPs, or land acquisition
Education programs and events	continuous	Composite measure of organizations conducting educational programs, outreach, events, or public engagement activities

6.3 Data Analysis

In this section, we introduce the dependent variable (DV) for our analysis: *perceived water quality improvement*. This variable captures whether stakeholders believe water quality in their watershed has improved since their organization became involved in watershed-related activities. To systematically examine the factors associated with these perceptions, we employ a quantitative modeling approach. Specifically, we use a binary response model (logit) to estimate the likelihood that stakeholders perceive water quality improvement, given a range of organizational, contextual, and demographic covariates. The following subsections describe the operationalization of the DV, the independent variables included in the analysis, and the results of both bivariate and multivariate statistical models.

6.3.1 Results for Univariate Statistics

Table 6.6 shows the summary statistics for all the variables associated with Tables 6.1 through 6.5. Looking at the dependent variable for perceived water quality improvement, stakeholders in 76% ($\hat{\sigma} = 0.43$) of the watersheds in this sample believe that quality is improving. We note that some improvements (based on perceptions) are from historical baselines. The Watershed Central New York noted "Water quality in Onondaga Lake has improved significantly over the past decade and especially compared to 40–50 years ago; Lake tributaries have improved somewhat in past decade, and significantly over past 40–50 years." Similarly, the Alliance of Rouge Communities noted that 23 years ago the entire Rouge failed to meet water quality standards 99% of the time. Now it meets standards (>98%). Others noted that improvements were due to the installation of BMPs. Cuyahoga River Community Planning Organization, for example, observed "There is less PCB and heavy metal contamination. Septic systems and manure management is improving, leading to a reduction in E.coli and nutrient contamination. BMPs on agricultural land have reduce[d] nitrate and phosphate runoff in addition to reducing sediment load."

This finding is consistent with the general observation that water quality is improving across the United States. There are several possibilities that likely underpin this general observation. First, water quality across the United States has improved. Today, many bodies of water are observably clearer and cleaner, potentially contributing to the widespread belief that water quality is improving, even if collaboration is largely unrelated to such successes. Second, stakeholders across a variety of watersheds have implemented visible best management practices and have produced a wide array of collaborative outputs including public engagement and clean up events. Third, watershed collaborations have also worked to educate the public and other stakeholders about various water quality initiatives,

Table 6.6 *Summary statistics for Phase Two data analysis ($n = 114$). Values reflect central tendencies and spread across key variables.*

Variable	Mean	Std Dev.	Min	Max
Dependent Variable				
Perceived water quality improvement	0.763	0.427	0	1
Stakeholder Organizations				
Local nongovernmental	6.781	5.989	0	32
Local governmental	2.505	3.797	0	17
Special natural resource-related districts	1.228	1.886	0	9
State nongovernmental	2.246	2.393	0	12
State governmental	2.693	2.239	0	10
Regional nongovernmental	0.763	1.306	0	6
Regional governmental	0.167	0.419	0	2
National nongovernmental	0.307	0.653	0	3
National governmental	1.974	2.067	0	9
University affiliate	0.965	1.545	0	7
Tribal governmental	0.281	0.747	0	4
Federated nongovernmental	1.158	1.202	0	5
Flow				
No natural flowing waters	0.140	0.340	0	1
Natural flow regime	0.693	0.463	0	1
Stakeholder Community Characteristics				
Watershed partnership	0.886	0.319	0	1
Political efficacy	3.750	0.894	1	5
Trust in governmental stakeholders	6.246	1.608	2	9.5
Trust in nongovernmental stakeholders	6.685	1.522	2	10
Collective Action Beliefs				
Problem diffusion	54.987	21.706	9	100
Natural process scientific understanding	5.261	1.765	1	10
Causal understanding of problems	5.373	1.883	1	10
Decision-making Process Legitimacy				
Decision-making fairness	3.263	0.845	1	5
Decision-making representativeness	3.641	0.786	1	5
Policy Output Characteristics				
Environmental improvement activities	3.185	1.418	0	9
Educative activities	5.272	2.163	0	11
Watershed Context and Conditions				
National estuary program (NEP) watershed	0.298	0.460	0	1
SES	0.171	0.084	0.051	0.392
Population density	252.103	572.366	0.583	4221.189
% White population	0.845	0.142	0.266	0.992
Watershed size (square miles)	1439.384	740.832	209.900	3633.200
% Interment flow	0.324	0.270	0	0.970
Prior appropriation doctrine	0.342	0.477	0	1
NPS pollution activism	14.794	7.916	2	25.500

improvement efforts, ongoing challenges, and raising awareness of water quality overall. Fourth, as compared to a few decades ago, the number of groups active in the environmental policy arena has proliferated. In this way, groups perceive improvement because they have a direct role in achieving the improvement.

6.3.1.1 Organizations within Watersheds

Moving to organizations involved in the sample of watersheds, Table 6.6 shows that an average of 6.78 ($\hat{\sigma} = 5.99$) local-level nongovernmental organizations are involved in watershed-related activities, ranging from 0 to 32. The average number of local-level governmental organizations participating in watershed-related activities is 2.51 ($\hat{\sigma} = 3.80$), ranging from 0 to 17. The average number of special districts involved in each watershed is 1.23 ($\hat{\sigma} = 1.89$), ranging from 0 to 9. Turning to the state and regional levels, the average number of state-level nongovernmental organizations is 2.25 ($\hat{\sigma} = 2.39$), ranging from 0 to 12, and the average number of state governmental organizations is 2.69 ($\hat{\sigma} = 2.24$), ranging from 0 to 10. The average number of regional nongovernmental organizations involved in watershed-related activities is 0.74 ($\hat{\sigma} = 1.31$), ranging from 0 to 6. The average number of regional governmental organizations is 0.17 ($\hat{\sigma} = 0.42$), ranging from 0 to 2.

At the national level, an average of 0.31 ($\hat{\sigma} = 0.56$) nongovernmental organizations are involved in this sample of watersheds, ranging from 0 to 3, and 1.97 ($\hat{\sigma} = 2.07$) governmental organizations are involved, ranging from 0 to 9. Nearly one ($\bar{x} = 0.97$; $\hat{\sigma} = 1.55$) university-affiliated organization is involved in watershed-related activities in the sample, ranging from 0 to 7. An average of 0.28 ($\hat{\sigma} = 0.75$) tribal government agencies are involved in sampled watersheds, ranging from 0 to 4. And, finally, an average of 1.16 ($\hat{\sigma} = 1.20$) federated organizations are involved, ranging from 0 to 5.

As for the remaining variables, 14 ($\hat{\sigma} = 0.34$) of the watersheds in the sample have no natural flowing waters, and a natural flow regime is in place in 69 ($\hat{\sigma} = 0.46$) of the watersheds. Stakeholders indicated that a watershed partnership is present in 89 ($\hat{\sigma} = 3.19$) of the sample. On a scale of one to five, stakeholders in the average watershed aggregately scored their political efficacy at 3.75 ($\hat{\sigma} = 0.89$). On a scale of 1 to 10, they scored their trust in governmental stakeholders at 6.25 ($\hat{\sigma} = 1.61$), while trust in nongovernmental stakeholders was slightly higher, at 6.69 ($\hat{\sigma} = 1.52$). For problem diffusion, stakeholders estimated that it would take 55 ($\hat{\sigma} = 21.71$) of stakeholders to change their behavior or activities to resolve watershed-related water quality problems. When asked about scientific understanding, stakeholders in the average watershed placed themselves roughly in the middle: 5.26 ($\hat{\sigma} = 1.77$) on a 10-point scale for understanding of natural processes, and 5.37 ($\hat{\sigma} = 1.89$) for understanding of causal processes. In terms of legitimacy perceptions, the average rating for fairness of the process was 3.26 ($\hat{\sigma}=0.85$), and for representativeness, 3.64 ($\hat{\sigma}=0.79$), both on 5-point scales. Finally, in the year prior to the survey, an average of 3.19 ($\hat{\sigma}=1.42$) environmental improvement activities and 5.27 ($\hat{\sigma}=2.16$) educative activities took place across sampled watersheds, with ranges from 0 to 9 and 0 to 11, respectively.

Beyond these variables, Table 6.6 describes several watershed context and condition variables. Among these is a variable for the NEP, which, as a dichotomous indicator, shows an average of 0.30 ($\hat{\sigma}=0.46$) watersheds in this sample are involved in the NEP. Socioeconomic status (SES) is measured as the percentage of households within the watershed with an annual income equal to or greater than $75,000 in 2000. This table also shows that an average of 17% ($\hat{\sigma}=0.08$) of watershed households have incomes at this level, although this number ranges from 5% to 39%. As for population density, the table shows there were an average of 252.10 persons per square mile ($\hat{\sigma}=572.37$) living in sampled watersheds in 2000; this number ranges from 0.58 to 4,221.19. Watersheds in the sample have an average white population of 85% ($\hat{\sigma}=0.14$), with a range of 27% to 99%. The average watershed is 1,439.38 square miles ($\hat{\sigma}=740.83$), with a minimum of 209.9 square miles and a maximum of 3,633.2 square miles. Apart from the NEP variable, these variables were gathered from the 2000 US Decennial Census.

Three additional variables measure ecological and civic conditions within the watershed. Sampled watersheds have an average 32% ($\hat{\sigma}=0.27$) intermittent flows; this ranges from no intermittent flow to nearly all intermittent flows (97%). A dichotomous indicator for the prior appropriation (Western) water doctrine shows that 34% ($\hat{\sigma}=0.48$) of sampled watersheds use this approach for determining water rights. Finally, the average NPS pollution activism score for sampled watersheds is 14.79 ($\hat{\sigma}=7.92$), ranging from 2 to 25.5 on a 0-to-30-point scale.[2] The discussion now turns to how these independent variables are correlated with the dependent variable of perceived water quality.

6.3.2 Bivariate Results

Table 6.7 shows there are several variables that have a statistically significant correlation with the perception that water quality has improved in the sampled watersheds.

The bivariate analysis shows that several variables are significantly correlated with perceptions of water quality improvement. As shown previously, a greater number of local nongovernmental organizations ($\hat{\rho}=0.222$; $p=0.017$) and local

[2] See Lubell et al. (2002) for the NPS activism measure.

Table 6.7 *Pairwise correlations ($\hat{\rho}$) and associated p-values for key organizational types in Phase Two of the study.*

Variable	Correlation ($\hat{\rho}$)	*p*-value
Local nongovernmental	0.222	0.018
Local governmental	0.226	0.016
Special districts	0.112	0.237
State nongovernmental	0.066	0.485
State governmental	0.173	0.065
Regional nongovernmental	0.121	0.201
Regional governmental	0.074	0.433
National nongovernmental	0.073	0.443

governmental organizations ($\hat{\rho} = 0.226$; $p = 0.016$) are both positively and significantly associated with stakeholder perceptions of improved water quality. Three stakeholder community characteristics also exhibit statistically significant correlations: the presence of a watershed partnership ($\hat{\rho} = 0.255$; $p = 0.006$), political efficacy ($\hat{\rho} = 0.234$; $p = 0.012$), and trust in governmental stakeholders ($\hat{\rho} = 0.100$; $p = 0.006$). Finally, natural scientific understanding – stakeholders' knowledge of the ecological processes within the watershed – is also positively correlated with perceived improvements in water quality ($\hat{\rho} = 0.165$; $p = 0.079$), though this association is only marginally significant.

Decision-making process legitimacy also matters to stakeholders: Both fairness ($\hat{\rho} = 0.215$, $p = 0.022$) and representativeness ($\hat{\rho} = 0.234$, $p = 0.012$) are significantly correlated with perceived water quality improvement. Among operational-level activities, a greater number of educative programs is also significantly associated with positive perceptions ($\hat{\rho} = 0.214$, $p = 0.022$). Several contextual and ecological variables exhibit significant correlations as well. These include participation in the NEP ($\hat{\rho} = 0.363$, $p = 0.000$), higher socioeconomic status ($\hat{\rho} = 0.195$, $p = 0.038$), greater population density ($\hat{\rho} = 0.191$, $p = 0.041$), and a higher proportion of white residents ($\hat{\rho} = 0.224$, $p = 0.016$). In contrast, greater levels of intermittent flow are negatively correlated with perceived improvement ($\hat{\rho} = -0.273$, $p = 0.003$). Finally, NPS policy activism is positively correlated with water quality perceptions ($\hat{\rho} = 0.194$, $p = 0.039$). The following section develops a multivariate model to assess the independent impact of these variables on stakeholder perceptions of water quality improvement.

As discussed previously, stakeholders' perceptions of water quality are shaped by a complex interplay of institutional, social, contextual, and environmental factors. These influences are further mediated by governance structures, targeted engagement efforts, and specific watershed characteristics. Table 6.8 outlines

Table 6.8 *Possible explanations for observed correlations between specific organizational characteristics and perceptions of water quality improvement.*

Factor	Possible reason
Greater number of local NGOs	Local NGOs often engage in advocacy, public education, and monitoring activities.
Greater number of local governmental organizations	Local governments may enforce water-related ordinances and visible policy action.

possible explanations for the observed correlations between organizational characteristics and perceptions of water quality improvement. For instance, a greater presence of local nongovernmental organizations (NGOs) may enhance perceptions of improvement because these organizations often lead advocacy, public education, and monitoring activities, thereby raising awareness and visibility of water quality initiatives. Likewise, a higher number of local governmental organizations can result in more effective enforcement of water-related policies and the implementation of visible actions, both of which may positively influence stakeholder perceptions.

6.3.3 A Binary Response Model

To analyze the independent effect of the significant variables on *perceived water quality improvement*, a binary response model (BRM) with a logit function was used. This approach is appropriate because it measures the proportional nature of the variable – a binary indicator of whether water quality has improved since stakeholder engagement in the watershed. Since this variable was created from aggregate responses for all those organizations responding for a given watershed, using this function allows the variable to range from $-\infty$ to ∞ when transformed into an odds ratio, as shown in the following equation.

$$\frac{\Pr(y = 1 \vee x)}{\Pr(y = 0 \vee x)} = \frac{\Pr(y = 1 \vee x)}{1 - \Pr(y = 1 \vee x)}$$

This odds ratio describes how often watershed stakeholders perceive the constellation of actors involved in their watershed is having an impact on the perception of perceived water quality improvement.

When initially creating a logit model with each of the variables having a significant impact, only two appear to be significant: population density and the percent of white population. A possible explanation is that several of the variables were having an equal effect on the dependent variable. Following Long (1997), several Wald tests were conducted to rule out this potential impact. First, it was possible that the effects of decision-making fairness and decision-making representativeness have an equal effect. This hypothesis was unable to be rejected at the $\alpha = 0.05$ level ($\chi^2 = 0.30$, $df = 1$, $p = 0.584$).

Second, it was conceivable that, due to the nature of the question, political efficacy may have an equal impact as decision-making representativeness.[3] Third, also due to the way the respondents answered the questions, political efficacy and decision-making fairness may also have an equal effect; this is also unable to be rejected at the $\alpha = 0.05$ level ($\chi^2 = 1.67$, $df = 1$, $p = 0.196$). Fourth, population density and percent white population may have an equal effect; this dynamic can be rejected at the $\alpha = 0.05$ level ($\chi^2 = 6.18$, $df = 1$, $p = 0.013$).

Finally, there is the possibility that political efficacy and watershed partnership may have an equal effect on one's perceptions of water quality outcomes. And, again, this relationship cannot be rejected at the $\alpha = 0.05$ level ($\chi^2 = 0.40$, $df = 1$, $p = 0.525$). This may indicate the need to examine various combinations of variables across several logit models.

Results from the logit models are presented in Table 6.9, each including a different combination of variables. Based on the Bayesian Information Criterion (BIC), Model 4 is preferred because it has the lowest BIC value. The strength of evidence for Model 4 is indicated by the difference in BIC between the two best models: 141.942 – 116.860 = 25.082. According to Long (1997), this magnitude of difference provides very strong support for Model 4.

Additionally, although not shown in the table, the Akaike Information Criterion (AIC) is also lowest for Model 4 (AIC = 105.915), further supporting its selection. Model 2 has the next lowest AIC (AIC = 109.089). The AIC differences are smaller than those for BIC, likely because BIC penalizes model complexity more heavily than AIC due to its identification assumptions (Long and Freese 2006).

Model 4 demonstrates that the presence of a watershed stakeholder partnership significantly increases the odds that stakeholders perceive water quality improvement in their watershed ($p < 0.05$). Perceptions of improvement are also more likely in watersheds with a higher percentage of white residents ($p < 0.05$) and in areas with greater population density ($p < 0.05$). Examining the watershed partnership variable more closely, the results indicate that the odds of perceiving water quality improvement rise substantially when a partnership is present.

6.4 Explanations

The results of the logit model indicate that the presence of a watershed partnership is a significant predictor of perceived water quality improvement. Specifically,

[3] This potential relationship was unable to be rejected at the $\alpha = 0.05$ level ($\chi^2 = 0.12$, $df = 1$, $p = 0.728$).

Table 6.9 *Logit models estimating perceived water quality improvement. Values shown are logit coefficients with standard errors in parentheses. Asterisks denote significance levels.*

Variable	Model 1	Model 2	Model 3	Model 4
Local Governmental Orgs.	−0.063 (0.073)	−0.045 (0.071)	−0.056 (0.073)	–
Local Nongovernmental Orgs.	0.157 (0.134)	0.159 (0.127)	0.170 (0.135)	–
Watershed Partnership	1.199 (0.934)	–	1.674* (0.874)	1.873** (0.750)
Political Efficacy	0.576 (0.358)	0.719* (0.335)	–	–
Trust in Governmental Orgs.	0.177 (0.177)	0.221 (0.174)	0.186 (0.171)	–
Natural Process Scientific Understanding	0.167 (0.157)	0.175 (0.157)	0.150 (0.150)	–
Educative Activities	0.135 (0.135)	0.106 (0.132)	0.225 (0.124)	–
SES	−5.956 (5.106)	−4.650 (4.904)	−6.350 (5.039)	–
Population Density	0.011* (0.005)	0.009* (0.005)	0.011* (0.005)	0.007* (0.003)
% White Population	4.759* (1.894)	4.816* (1.899)	4.580* (1.874)	4.286** (1.663)
% Intermittent Flow	−0.013 (0.012)	−0.011 (0.012)	−0.011 (0.012)	–
NPS Policy Activism	0.013 (0.037)	0.013 (0.037)	0.013 (0.036)	–
Intercept	−8.171** (2.500)	−8.158*** (2.448)	−6.778** (2.258)	−4.789** (1.641)
Model Fit				
BIC	144.973	141.942	142.942	116.860
LR χ^2	41.41***	39.27***	38.70***	26.90***

Note: Standard errors in parentheses. $^{*}p < 0.05$, $^{**}p < 0.01$, $^{***}p < 0.001$. n=114.

stakeholders in watersheds with partnerships are approximately six times more likely to perceive improvements in water quality compared to those without partnerships. This finding underscores the importance of collaborative governance structures in shaping stakeholder perceptions and highlights the role of partnerships in facilitating effective water management practices.

This section interprets the key findings from the quantitative analysis of perceived water quality improvement among watershed stakeholders. By examining the relationships between organizational, demographic, and contextual variables and stakeholders' perceptions, we can better understand the factors that shape optimism or skepticism about collaborative water management efforts. The discussion highlights the implications of these results for both theory and practice, considers the limitations of the current analysis, and suggests directions for future research. In particular, we focus on the roles of demographic context and watershed partnerships in shaping perceptions, and reflect on the broader significance of these findings for collaborative governance and environmental policy.

6.4.1 Demographics Again Matter?

Demographic factors are associated with how stakeholders perceive water quality. Here, stakeholders in watersheds with higher percentages of white residents are associated with higher perceptions of water quality. Similar to results related to revealed water quality, this dynamic may be linked to historical patterns related to infrastructure, historical water quality, and development planning. Watersheds located within areas with a larger white population, for example, may have benefitted from greater investment in water-related infrastructure, more robust regulatory enforcement, and more access to (and responsiveness from) state and local policymakers relative to water quality as compared to disadvantaged communities. This finding aligns well with the environmental justice literature that has noted links between racial and ethnic minorities, lower SES, and higher rates of exposure to environmental hazards, such as air and water contamination. These historical patterns matter a great deal because improving water quality can be costly and take years to complete. In short, in watersheds with more racially homogenous populations, individuals may not only enjoy better water quality, but also believe they receive better water quality.

Higher population densities were also connected to improved perceptions of water quality. Several possibilities may help make sense of this finding. First, areas with a greater number of individuals may receive increased environmental monitoring, have heightened levels of public awareness, and more resources for oversight as compared to less populated areas. It is also possible that in these areas there are more local governments, and thus more resources to maintain, promote, and communicate water quality improvements, which are all outputs that relate to visibility and actions that demonstrate "something" is being done. Additionally, these areas may also have a greater concentration of advocacy organizations, which can support education activities such as water conservation programs.

6.4.2 Watershed Partnerships in Perceived Water Quality

One of the common goals of watershed partnerships is to support multi-sector collaboration. This is strongly supported by the results in Table 6.9. In Model 4, the coefficient for the partnership variable is 1.873 (SE = 0.750, $p < 0.01$), which translates to an odds ratio of $\exp(1.873) \approx 6.51$. In other words, the presence of a watershed partnership is associated with more than six times higher odds of perceiving improvement in water quality compared to watersheds without partnerships. This finding is intuitive: Partnerships are often linked to visible and tangible outputs such as education, restoration efforts, and the promotion of BMPs. Furthermore, perceived effectiveness is likely mediated by the visibility of these efforts, stakeholders' prior knowledge about water quality and the watershed, and specific communication and information-sharing strategies, all of which partnerships can help facilitate.

A possible explanation for this finding relates to the reality that many of the stakeholders engaged in watershed partnerships are keenly aware of the current and future challenges confronting their watershed, relevant resource limitations, and other constraints related to policy. These factors, in turn, contribute to respondents diving more deeply into producing collaborative outputs, sharing information, and engaging with other stakeholders. The production of collaborative outputs related to a partnership may further support perceived improvement. As BMPs are installed, even if it may take several years for water quality to show signs of improvement, they are visible reminders of progress and action. And, even without immediate results, they may contribute to other goals such as protecting the watershed (rather than restoring it) or cleaning a small segment, potentially leading to perceived improvements. Finally, it may be that stakeholder perceptions are influenced by communication and engagement strategies that focus on loss or the ecological harms threatening water system health. This dynamic may also increase the urgency for action and beliefs that collaboration is necessary.

6.5 Conclusion

The analyses presented here provide tentative support for our model. The bivariate relationship analysis indicates relationships between local nongovernmental and governmental organizations, respectively, and perceptions about water quality improvement. The presence of a watershed stakeholder partnership, political efficacy, collective action beliefs, and trust in governmental organizations were also associated with higher perceptions of collaboration effectiveness and goal attainment. With regard to decision-making process legitimacy, there is an association among fairness and representativeness with perceived water quality. Finally, there is support related to operational-level educative activities. Of course, it must be mentioned that among these, only the relationship relative to watershed stakeholder partnership found support in the multivariate BRM.

While the preceding section outlines key statistical relationships, perceptions are not merely reflections of institutional or environmental variables. Instead, they are shaped by broader dynamics, such as community histories, stakeholder roles, and the motivations that drive participation. The following section situates these findings within a wider context, highlighting how individual and collective beliefs about progress interact with the structure and visibility of collaboration.

6.5.1 Reflections on Perception, Optimism, and Collaborative Value

Our analyses suggest that there is a mismatch between the perceptions of collaboration participants and actual (measurable) changes in water quality. In general, collaboration participants tend to have very positive views of their efforts and are more likely to hold optimistic impressions of the work of the collaboration. Areas that are whiter and more densely populated tend to hold the most optimistic views of their work as compared to more diverse communities. This overall finding is consistent with similar findings in the literature (Parisi et al. 2004; see also Morris et al. 2013).

Our focus is on existing collaborations; all our respondents are part of ongoing collaborative efforts. It seems self-evident that people have choices about how they spend their time and the things to which they apply their efforts, and most are unlikely to put in much effort if there is nothing to show for it. People need to believe their efforts are making a difference. Put another way, people want to be associated with a successful effort, and optimism helps validate that association. However, this optimism is not a given and can be upended by several factors, such as the quality and veracity of available information and data. In the case of watershed management, this can lead to dissatisfied stakeholders leaving the collaborative effort.

As of today, water quality data are imperfect. Data points are often of poor quality, and the less-than-optimal mechanisms for reporting water quality mean the data are often incomplete. Other research has noted the many challenges of using reported water quality data (see, e.g., Aistrup and Morris 2024; Mayer et al. 2024). In short, the reporting is infrequent, often incomplete, and arbitrary. There is anecdotal evidence that some states are vigilant about careful monitoring practices (see, Morris et al. 2024), but the data states report to EPA are often plagued with a variety of problems. It is possible that there is more improvement in each watershed than is evident in the reported data. Nonetheless, the EPA data remain the best available source that can be applied at the national level. A future study might involve collecting data from state environmental or health departments for specific watersheds, although collecting such data – even for a cross-sectional study – would require significant resources.

Another point to consider is that, while perceived versus measured changes in water quality are useful indicators, collaboration has benefits that move well beyond improved water quality. There is a substantial literature that suggests increased social capital is an important output of the collaboration process (Morris et al. 2013; Sabatier et al. 2005; Putnam 2000). In other words, to think of watershed collaboration in a single dimension – water quality – likely misses much of the benefit of the effort and at least part of the reason for collaboration in the first place. Diaz-Kope and Morris (2019) explore the differences in motivation between organizational participants from the public, private, and nonprofit sectors; they report a variety of different motivations driven by sector. However, a common thread is the desire to be part of something in the community. Diaz-Kope et al. (2024) note that a commitment to place often drives individuals to join watershed collaborative efforts; this is very much a community-driven motivation. Finally, as Gray (1985) and Wood and Gray (1991) note, the point of collaboration is to tackle intractable problems. Wicked problems, by their very nature, are complex and interconnected, without easy solutions. Collaboration can still have impact, even if that impact is not reflected in water quality measurements.

References

Aistrup, Joseph A., and Morris, John C. 2024. Does Institutional Location Matter? The Effect of Agency Type on Resource Distribution for Nonpoint Water Pollution. *Water Policy*, **26**(10), 1039–1055.

Balazs, Carolina L. 2011. *Just Water? Social Disparities and Drinking Water Quality in California's San Joaquin Valley*. Ph.D. thesis, University of California, Berkeley.

Balazs, Carolina L., and Ray, Isha. 2014. The Drinking Water Disparities Framework: On the Origins and Persistence of Inequities in Exposure. *American Journal of Public Health*, **104**(4), 603–611.

Bowen, William. 2002. An Analytical Review of Environmental Justice Research: What Do We Really Know? *Environmental Management*, **29**(1), 3–15.

Bullard, Robert D. 2018. *Dumping in Dixie: Race, Class, and Environmental Quality*. Routledge.

Collins, Timothy W., and Grineski, Sara E. 2017. Environmental Justice and Flood Hazards: A Conceptual Framework Applied to Emerging Findings and Future Research Needs. In *The Routledge Handbook of Environmental Justice*, 350–361. Routledge.

Collins, Timothy W., Grineski, Sara E., Chakraborty, Jayajit, and Flores, Aaron B. 2019. Environmental Injustice and Hurricane Harvey: A Household-Level Study of Socially Disparate Flood Exposures in Greater Houston, Texas, USA. *Environmental Research*, **179**, 108772.

Diaz-Kope, Luisa, and Morris, John C. 2019. *Organizational Motivation for Collaboration: Theory and Evidence*. Lanham, MD: Lexington Press.

Diaz-Kope, Luisa, Krawczyk, Kelly, Morris, John C., and Weber, Sindhu. 2024. Commitment to Place as Motivation for Citizen-Based Watershed Collaboration: The "BIMBY" Effect. *Journal of Public and Nonprofit Affairs*, **10**(2), 221–238.

Gray, Barbara. 1985. Conditions Facilitating Interorganizational Collaboration. *Human Relations*, **38**(10), 911–936.

Imperial, Mark T. 2005. Using Collaboration as a Governance Strategy: Lessons from Six Watershed Management Programs. *Administration & Society*, **37**(3), 281–320.

Long, J. Scott, and Freese, Jeremy. 2006. *Regression Models for Categorical Dependent Variables Using Stata*. Vol. 7. Stata Press.

Long, J. Scott, and Long, John Scott. 1997. *Regression Models for Categorical and Limited Dependent Variables*. Advanced Quantitative Techniques in the Social Sciences, vol. 7.

Lubell, Mark. 2005. Do Watershed Partnerships Enhance Beliefs Conducive to Collective Action? In *Swimming Upstream: Collaborative Approaches to Watershed Management*, edited by Paul A. Sabatier, Will Focht, Mark Lubell, Zev Trachtenberg, Arnold Vedlitz and Marty Matlock, 201–232. Cambridge, MA: The MIT Press.

Mayer, Martin K., Morris, John C., Zhang, Xiaodan, and McNamara, Madeleine W. 2024. Explaining State Efforts to Create TMDL (Total Maximum Daily Load) Agreements. *Social Science Quarterly*, **105**(5), 1776–1790.

Mohai, Paul, Pellow, David, and Roberts, J. Timmons. 2009. Environmental Justice. *Annual Review of Environment and Resources*, **34**(1), 405–430.

Morris, John C., Gibson, William A., Leavitt, William M., and Jones, Shana C. 2013. *The Case for Grassroots Collaboration: Social Capital and Ecosystem Restoration at the Local Level*. Lanham, MD: Lexington Books.

Ostrom, Elinor. 2005. Policies that Crowd Out Reciprocity and Collective Action. In *Moral Sentiments and Material Interests: The Foundations of Cooperation in Economic Life*, 253–275.

Parisi, Domenico, Taquino, Michael, Grice, Steven M., and Gill, Duane A. 2004. Civic Responsibility and the Environment: Linking Local Conditions to Community Environmental Activeness. *Society and Natural Resources*, **17**(1), 97–112.

Putnam, Robert D. 2000. *Bowling Alone: The Collapse and Revival of American Community*. Simon and Schuster.

Sabatier, Paul A., Focht, Will, Lubell, Mark, Trachtenberg, Zev, Vedlitz, Arnold, and Matlock, Marty (eds.). 2005. *Swimming Upstream: Collaborative Approaches to Watershed Management*. Cambridge, MA: The MIT Press.

Walker, Gordon. 2012. *Environmental Justice: Concepts, Evidence and Politics*. Routledge.

Wood, Donna, and Gray, Barbara. 1991. Towards a Comprehensive Theory of Collaboration. *Journal of Applied Behavioral Science*, **27**, 139–162.

Zhao, Jing, Gladson, Laura, and Cromar, Kevin. 2018. A Novel Environmental Justice Indicator for Managing Local Air Pollution. *International Journal of Environmental Research and Public Health*, **15**(6), 1260.

7

Finding Voices from the Field: Insights from Stakeholders

Whether collaboration, as a set of activities and interactions, matters depends largely on the goals of the actors involved. Commonly cited goals within watersheds include enhancing ecological health and overall sustainability, improving the quality of life for stakeholders, reaching intended and designated uses, building and maintaining social capital and trust, and capitalizing on community and economic development opportunities. Stakeholders may strengthen relationships among interested groups but have more muted success in improving watershed health, as shown in Chapter 6. In other words, reaching one or more of these possible goals is not guaranteed. In fact, achieving any of the goals associated with a watershed, as well as sustaining the relationships inherent to collaboration, requires a robust understanding of the history and challenges encountered by diverse stakeholders. Given the complexity of many collaborations, challenges may relate to processes, relationships, outputs, and their influence on outcomes. This chapter presents a detailed narrative based on survey responses gathered from a range of watershed stakeholders who are actively working within a watershed collaboration. To illustrate key themes, the chapter draws on both survey data and open-ended responses, and is structured around many of the core organizing concepts presented in our model.

7.1 Values and Goals

We begin by unpacking many of the reported goals and priorities of watershed collaboration stakeholders. In general, respondents held a variety of priorities, which ranged from mitigating the effects of pollution and climate change to managing the pressures associated with poor land use planning and rapid growth.

Explicating the presence of so many diverse values makes sense when considered alongside the explanatory model presented in Chapter 3:

- The external environment, characterized by political factors, limited funding, competing priorities, and societal pressures, may facilitate or impede collaboration and related watershed goals and values. In either case, the external environment can influence how goals are set and pursued.
- The watershed's context, including its ecological and civic conditions, frames which goals are possible, legitimate, and salient, and which are viewed as achievable.
- Information communities facilitate the sharing of knowledge and experiences related to values and goals, and each can be (or has been) translated into action.
- Stakeholder characteristics can support a goal or set of goals and values, or they can contribute to conflict.
- Legitimacy is associated with the collaboration process and whether or not stakeholders believe the goals are fair, attainable, and consistent with their values.

The goals of watershed collaboration, we suggest, are directly tied to the external environment in which the collaborative group operates. We also note that the development of those goals is likely shaped by contextual factors as well as the civic and ecological conditions present in the watershed. The watershed's history, for example, likely influences present efforts, particularly if the collaboration is seeking to protect or restore the watershed. The composition of the stakeholder community is also likely to influence goals, with characteristics such as previous experiences, openness to learning, and stakeholder needs shaping what is possible and what is realistic. Cumulatively, these goals help explain the range of collaborative outputs observed in Chapter 5, as well as the perceptions of collaboration outcomes reported in Chapter 6.

Table 7.1 offers a snapshot of stakeholder reflections with an emphasis on goals and priorities. Taken together, they offer insight into some of the most pressing challenges that watershed collaborations are confronting. Importantly, they also represent different contexts and voices, including those from conservation districts and regional partnerships.

Within our survey, federated organizations and regional NGOs demonstrate the strongest support for collaborative watershed management. Their estimates indicate a strong belief that a high percentage of stakeholder participation is necessary to achieve meaningful environmental outcomes and relevant goals. This perspective appears to stem from their extensive experience with the interconnected nature of watershed systems and their understanding of how actions in one area influence broader ecological conditions. Their emphasis on broad stakeholder involvement likely reflects direct observations of how water quality issues often transcend jurisdictional and organizational boundaries.

Table 7.1 *Examples of watershed collaboration challenges where shared objectives are contested or difficult to maintain, due to environmental, political, or cultural pressures.*

Watershed collaboration	Example observations – greatest challenge?
Orange County Water Authority	Balancing competing land uses (agricultural, conventional development . . .) in a setting impacted by disruptive storm events including Hurricanes Irene and Lee, Superstorm Sandy, and various ice, snow, and microburst rain events.
N/A	Increasing severity of droughts and floods, attributed to climate change; weak enforcement from the Alabama Department of Environmental Management; poor forestry and agricultural practices; and absence of a comprehensive watershed management plan with stakeholder input and action timeline.
Belknap County Conservation District	Funding cuts across government services have weakened resource management, education, and infrastructure. Tensions arise between supporting individual landowners and public accountability for spending, undermining multi-year project stability.
Elk County Conservation District	Industry pressure, economic decline, limited public understanding, and ineffective regulation all complicate watershed management efforts.
N/A	Undermined agencies, deregulation efforts, and uncertain efficacy of TMDL modeling raise doubts about the state's ability to manage climate impacts and achieve lasting change.
N/A	Anticipated conflict over water use for fracking and its downstream effects. Public misunderstandings about flood mitigation (e.g., believing tile drainage reduces flooding) hinder sustainable practices.
N/A	A lingering belief that undeveloped land is wasteful, resistance to conservation, and divergent understandings of "livability" and "sustainability" create deep divides in stakeholder values and visions.

Source: Generated by authors.

Stakeholders in more agricultural areas, for example, explain that reaching their goals would likely involve upstream agricultural practices even if they are located downstream. In other watersheds, stakeholders note how urban development impacts habitats, and how individual and collective behaviors influence watershed health. This accumulated experience informs their understanding of the systemic nature of environmental impacts. Their comprehensive view of watershed systems enables them to identify the factors influencing environmental outcomes and emphasize the importance of integrated governance approaches and coordinated action among diverse stakeholders. Finally, their experience suggests that meaningful environmental improvements require sustained collaboration across communities and sectors.

Other stakeholder groups offer alternative perspectives. Local agencies present a particularly notable case. Despite operating independently, they frequently share perspectives with their partnered counterparts regarding the necessity of broad stakeholder participation. This alignment likely stems from their direct engagement with community-level environmental issues, where they regularly observe how stakeholder actions affect watershed conditions. Additionally, their position enables them to understand the complexity of watershed systems and the importance of community engagement, even without formal partnership structures. International NGOs, however, demonstrate a markedly different perspective. Their lower estimates of required stakeholder participation suggest a more selective approach to watershed challenges. This viewpoint may reflect their broader operational scope and strategic priorities, potentially overlooking the significance of localized, community-level dynamics in maintaining watershed health.

These divergent perspectives reveal fundamental differences in how organizations conceptualize watershed challenges and goals. Organizations with primarily local experience tend to view watershed management as a complex system requiring multiple interventions and widespread behavioral changes. In contrast, organizations with a broader scope often approach these challenges through more targeted interventions, potentially underestimating the importance of comprehensive stakeholder engagement.

The distinction becomes especially evident when examining organizations' experiences with coordinated environmental actions. Organizations with limited involvement in collective efforts often approach watershed management independently. Their focus typically remains on immediate environmental concerns, potentially leading to an underestimation of the broader ecological interdependencies within watershed systems. This narrower perspective highlights the limitations of isolated approaches and the potential advantages of developing more collaborative frameworks for addressing shared environmental objectives.

7.2 Operational Choices and Outputs

Sabatier and colleagues (2005, 179) acknowledge that defining rules is difficult, as there is often a byzantine array of agencies, governments, statutes, and rules applicable to a particular geographic area or watershed. Recognizing this limitation, the authors emphasize evaluating the effectiveness of collaboration by whether actors "put projects on the ground and engage in other behavior that ultimately improve watershed health" (Sabatier et al. 2005, 179). This shift places emphasis on operational-level activities and sets the stage for Imperial's (2005, 289) observation that the "world of action occurs at the operational level." At this level, outputs and actors are often designed to (1) directly improve environmental conditions; (2) educate the public and decision makers; and (3) monitor conditions and enforce regulations. Activities include best management practices (BMP) installation, habitat restoration projects, educational programs for public officials, school curriculum development, and data collection on environmental activities, among several others. If watershed management programs – including stakeholder partnerships – are to improve environmental conditions, it will be through these operational-level activities (Imperial 2005; Sabatier et al. 2005).

These collaborative outputs are shaped by several important variables present in our model. The external environment is likely to influence the availability of resources, the saliency of watershed restoration or protection, and the degree to which there is political support for collaboration and its associated activities. Additionally, the context of the watershed, including its ecological health, impacts the needs of the watershed and the capacities of various stakeholders to address those ecological conditions.

Within this broader context, information communities offer opportunities for stakeholders to coordinate activities, establish trust and legitimacy, and promote adaptive learning and share experiences. The characteristics of stakeholders, including group composition, scale, and past collaboration experience, shape the organization's design, its beliefs and capacities, and operational activities in pursuit of goals. Finally, legitimacy reflects whether stakeholders trust one another, the collaborative process and decisions, and whether actors view outputs as necessary and supportive of overarching goals.

Operational choices are made within this broader organizational context and encompass the day-to-day implementation activities that directly impact environmental conditions, both real and perceived. Within watersheds, organizations routinely engage in water quality monitoring, restoration projects, and enforcement of regulatory requirements, all guided by established plans and protocols. These operational activities serve as the mechanisms through which higher-level values and constitutional constraints are translated into tangible actions that shape watershed health and the daily experiences of stakeholders. Table 7.2 summarizes these key activities:

As noted previously, watershed collaboration efforts involve a mix of education, management, proactive, and restoration activities. For example, respondents highlight efforts associated with public awareness campaigns, improvements related to municipal stormwater regulations, and the implementation of agricultural BMPs. According to respondents, these operations contribute to changes in individual behaviors, reductions in sediment and nutrient runoff, and the adoption of specific best management practices. While these activities represent important operational

Table 7.2 *Highlights of operational activities implemented by watershed collaborations to improve water quality, reduce erosion, and promote behavioral change.*

Watershed collaboration	Common activities
Adams County Office for Planning and Development	Awareness and education campaigns have influenced individual behaviors regarding yard care, littering, and pet waste. Municipalities have adopted stormwater regulations that reduced sediment loads, while agricultural BMPs helped lower nutrient runoff. A county-wide signage initiative now labels all streams along state roads.
Friends of Waterford Park	Educating the public on the limits of small-scale planting for erosion control has been a key focus. Despite progress, long grant cycles delay larger mitigation efforts, and some stream-adjacent trees may collapse before stabilization measures are implemented.
Skyline Ridge Neighbors	Portland eliminated combined sewer overflows from the watershed, significantly improving water quality – especially during wet seasons. Additional efforts include riparian restoration and instream habitat improvements.
Cuyahoga River Community Planning Organization	Contamination from PCBs and heavy metals has decreased. Improvements in septic and manure management have reduced *Escherichia coli* and nutrient pollution. Agricultural BMPs have lessened runoff of nitrates, phosphates, and sediment.

Source: Generated by authors.

efforts, they do not encompass work that requires larger-scale mitigation and restoration.

7.2.1 Document Production and Sharing Patterns

As noted previously, watershed collaboration efforts involve a mix of education, management, proactive, and restoration activities. For example, respondents highlight efforts associated with public awareness campaigns, improvements associated with municipal stormwater regulations, and the implementation of agricultural BMPs. These operations contribute (at least based on respondents' beliefs) to changes in individual behaviors, to reductions in sediment and nutrient runoff, and to the adoption of specific BMPs. While the observations above point to important operations, they also do not incorporate work that requires larger-scale mitigation and restoration work.

Local governmental agencies and local nongovernmental organizations (NGOs) demonstrate significant engagement, albeit with a more specialized focus. These organizations tend to concentrate on producing documents related to land use planning and watershed management plans, reflecting a strategic emphasis on spatial and environmental governance. Their more limited involvement in other document types suggests a targeted or capacity-driven approach, focusing on areas where their regulatory and planning tools are most effective. Regional NGOs, in particular, stand out as active contributors. Their participation is especially evident in the development of MOUs, watershed plans, and legislative proposals. Rather than isolated efforts, this pattern reflects a consistent role across multiple facets of governance. In this capacity, regional NGOs serve as intermediaries that help formalize agreements and advance both planning and policy agendas. Their sustained involvement points to the importance of these organizations in aligning community priorities with institutional frameworks.

By comparison, international NGOs and national governmental organizations exhibit a moderate level of engagement across a variety of governance documents, with particular emphasis on MOUs and watershed plans. This pattern suggests these organizations leverage their broader scope and influence to support high-impact governance initiatives, albeit with less intensive involvement than regional stakeholders. Special districts display the most limited participation. Their contributions are primarily restricted to watershed plans and legislative documents, indicating a narrow role in formalizing agreements and shaping governance structures. This constrained engagement reflects their specialized mandates and jurisdictional limitations. Regional NGOs are consistently more active across governance mechanisms, particularly in collaborative planning and implementation efforts.

The consistent pattern of higher document production among organizations, especially local governments and regional NGOs, engaged in partnerships reinforces the importance of collaboration in facilitating formal governance processes. Governance documents, far from being mere administrative artifacts, represent tangible commitments to structured and transparent approaches to watershed management. These findings emphasize the critical role of partnerships in establishing the institutional frameworks necessary for effective and sustainable watershed governance.

7.2.2 Focal Areas/Expertise

Our analysis of organizations participating in watershed partnerships reveals distinct patterns of engagement and focal areas. These patterns highlight how various organizational types differ in their involvement, relationship-building, and strategic approaches to watershed management. Such differences reflect the diverse institutional capacities, expertise, and priorities that shape each organization's contributions to collaborative efforts.

Local governmental agencies and local NGOs showed significant but specialized involvement in watershed partnerships. These organizations concentrate their engagement in land use planning and watershed management plans, leveraging their expertise in spatial and land use governance. Their more limited participation reflects resource constraints and jurisdiction-specific priorities that shape their institutional focus.

State governmental agencies and regional NGOs are highly involved. Both types are instrumental in establishing habitat restoration priorities, producing environmental progress reports, developing water quality assessments, and establishing formal shared goals. This engagement pattern demonstrates how partnerships enable organizations to expand beyond traditional operational roles into broader environmental governance activities. Regional NGOs also demonstrate the highest participation in developing memoranda of understanding, watershed management plans, and legislative contributions. This engagement pattern aligns with Agranoff's (2012) concept of "conductive organizations" that facilitate stakeholder collaboration and coordination.

International NGOs and national governmental organizations focus on specific high-impact activities. Their participation centers on developing MOUs and watershed management plans, suggesting a deliberate approach to maintaining presence across multiple watershed contexts. This targeted engagement allows these organizations to maximize their influence while operating within resource constraints. Special districts also demonstrate highly specialized engagement, focused on watershed plans and legislative activities. This narrow focus reflects their specific institutional mandates centered on infrastructure and service delivery within defined geographic areas. Their participation pattern indicates how organizational mission shapes collaborative engagement even within partnership frameworks.

Stemming from their practical experience in collaborative watershed management, federated organizations and regional NGOs also exhibit high levels of stakeholder involvement. Their work demonstrates that addressing complex environmental challenges requires moving beyond isolated interventions to embrace comprehensive approaches that engage multiple stakeholders. This recognition of the limitations of unilateral action positions them as strong advocates for collaborative solutions that bring together agricultural, urban, commercial, and environmental interests.

Similarly, national NGOs engaged in partnerships typically implement comprehensive watershed management strategies that span entire watersheds. These organizations coordinate multiple simultaneous initiatives, including restoration projects, volunteer programs, and conservation efforts. One organizational leader stated, "We've learned that piecemeal solutions don't work, we need to look at the whole system." This contrasts markedly with nonpartnering organizations, which typically implement targeted interventions focused on specific outcomes like habitat improvement or localized cleanup projects, reflecting their more constrained operational scope.

Examination of daily operations and challenges in watershed management organizations reveals distinct approaches shaped by organizational structure, partnerships, and resource availability. Survey responses and stakeholder interviews indicate these factors fundamentally influence operational strategies and implementation effectiveness. The differences manifest most clearly when comparing partnership and nonpartnership organizations across multiple operational dimensions.

7.2.3 Non-collaborating Organizations

The picture of engagement among non-collaborating organizations in watershed management reveals a striking contrast to their collaborating counterparts. Analysis of survey responses shows consistently lower levels of engagement in constitutional choice actions, though significant variation still emerges across different types of organizations. The reduction in activity is particularly pronounced for international NGOs and local NGOs. These organizations show notably lower participation across most governance activities. Despite this, some bright spots remain. Both groups demonstrate relatively higher levels of engagement in focused areas, such as setting priorities for habitat restoration and reporting progress toward environmental goals. This selective participation suggests that these organizations, even outside formal partnerships, are guided by independent mandates or strategic priorities, allowing them to contribute meaningfully to specific policy initiatives.

Federated organizations present an intriguing case. Despite their broad mandates, their involvement in actions like MOUs and watershed plans is significantly diminished without partnerships. This diminished engagement points to the vital role that partnerships play in enabling these organizations to participate fully in governance processes. Without the structure and resources provided by formal collaborations, federated organizations appear to struggle to sustain comprehensive involvement in watershed management activities.

Regional NGOs stand out as an exception to the overall trend, maintaining a high level of engagement even without formal partnerships. These organizations exhibit consistent activity, particularly in developing MOUs and watershed plans. This resilience underscores the capacity of regional NGOs to operate effectively and independently, driven by their strong local knowledge, community connections, and organizational adaptability. Their continued involvement suggests that regional NGOs can serve as drivers of governance processes, even in the absence of formalized collaborations.

In contrast, local government agencies and NGOs, which typically thrive within partnerships, experience a sharp decline in engagement when operating independently. Their activity levels are especially diminished in areas such as developing legislation and watershed plans. This reduction likely stems from several factors: limited resources, constrained capacity for independent action, and reduced access to collaborative networks and shared resources. The absence of supporting infrastructure and formalized opportunities for engagement further compounds these challenges, narrowing their focus and reach.

State governmental agencies, however, display a more robust pattern of independent involvement. These agencies sustain higher levels of engagement across a broad range of activities, notably in setting priorities for habitat restoration, producing reports on water quality enhancements, and monitoring progress toward environmental goals. This resilience reflects their statutory authority and resource availability, enabling them to lead policy development and implementation efforts even without partnership support.

Special districts, on the other hand, continue their characteristic pattern of limited engagement. Their involvement is primarily concentrated on infrastructure investments and formal monitoring protocols. This narrow focus underscores the extent to which these entities rely on partnerships to drive broader policy development and governance processes, constrained by their specialized mandates and geographic limitations.

The analysis of non-collaborating organizations offers valuable insights into watershed management efforts outside formal partnerships. Most organizations exhibit more focused, specialized engagement rather than comprehensive participation, with resource constraints and capacity limitations significantly shaping their activities. While some, particularly regional NGOs and state agencies, maintain effective independent operations, the absence of formal partnerships generally leads to fragmented and limited approaches to watershed governance. These findings highlight the critical importance of partnerships in fostering holistic and coordinated management strategies, even as some entities demonstrate the capacity to navigate independent pathways.

7.2.4 Learning, Relationships, and Institutional Trust

Watershed stakeholders are diverse. They include power companies that operate hydroelectric power-generating dams, homeowners who live in a watershed, utilities who provide water to communities, fisheries, industries, and governments at various levels, conservation groups, recreational users, agricultural users, sporting groups, academic researchers, and specialist practitioners. Sabatier and colleagues (2005, 20) remind us that stakeholders may also include a range of policymakers and experts who hold a variety of preferences and often respond to a multitude of incentives. In some instances, traditional battle lines are evident, and in others they break down as former adversaries become partners (Kash 2008). Thus, effective collaboration is not automatic; it often requires stakeholders to learn and trust one another. If relationships are positive and include trust, collaborations are likely better prepared to withstand shocks and to share information.

The ability and/or willingness to trust and learn are, in turn, shaped by several interrelated factors. External influences, such as funding changes or shifts in political leadership, can support or strain relationships within a collaboration by presenting new opportunities and constraints. Additionally, actors' context, such as their beliefs relative to the urgency and severity of the problem, the value of information and partners, and their beliefs about what is likely and possible, shapes their willingness to participate in a collaborative effort.

Information-sharing practices exemplify the importance of unpacking contextual conditions. If stakeholders believe that data are valuable and accurate, and in the integrity of their collaborators, trust and additional information are more likely. Additionally, the stakeholders themselves impact levels of learning and trust. As mentioned, collaborations often involve actors with varied interests, values, goals, and past experiences with one another. In some cases these factors align well; in others they are best described as disparate. Either can hinder or enhance collaboration. Finally, when stakeholders perceive decision-making as equitable and legitimate, trust is reinforced, enabling more durable and healthy collaborations.

Developing trust among stakeholders involved in the watershed contributes to successful collective action (Lubell 2007) if commitment is of primary concern (Scholz, Berardo, and Kile 2008). This challenge is noteworthy, as Bidwell and Ryan (2006) note that diverse stakeholder participation in watershed partnerships is often necessary to achieve changes in the management of the water resource, though more homogeneous or smaller groups may be more likely to conduct watershed improvement activities. Stakeholders described the importance of trust in Table 7.3.

Some respondents, albeit limited in number, highlight that effective watershed management is often associated with social dimensions such as the presence of long-term relationships, high levels of trust, and community engagement. While quick wins are often celebrated and encouraged (and required for funding!), when it comes to watershed health, stakeholders point to the importance of

Table 7.3 *Examples of how watershed collaborations build trust and cultivate lasting relationships with stakeholders to support environmental outcomes.*

Collaborative	Relationship building (as examples)
The Wetlands Conservancy	Successful collaboration depends on sustained trust, which takes years to build. Many achievements stem from over seven years of relationship-building. Yet, regulatory and funding frameworks often prioritize quick turnarounds and short-term metrics – misaligned with the time and flexibility needed for genuine collaborative success.
Friends of the Crooked River	Outreach must reach both local residents and transient tourists. As a seasonal destination, the area sees an influx of visitors who lack a lasting connection to the watershed yet significantly impact its water quality.
Cornell Cooperative Extension	Notable behavioral and environmental improvements include fewer crop fields near rivers, reduced cattle access to streams, better sewage controls, and shifting local attitudes. It's now socially unacceptable to dump trash in rivers – reflecting cultural change as much as policy enforcement.
Goose Lake Association	The challenge lies in fostering collective understanding of why change is necessary – achieving buy-in without imposing mandates.

Source: Generated by authors.

relationship-building and adaptive management, which can take years to build. Educating the public, for example (especially tourists or individuals without a strong connection to the region) remains a challenge. However, they point to indicators of progress. This includes changes in public attitudes, reductions in point-source pollution, and improved land use practices. In short, they point to proactive activities that prevent further ecological damage, even if they may not repair legacy pollution.

7.2.4.1 Trust in Stakeholders

Respondents provide additional insights into their perceptions of the stakeholder community, focusing on collective action, scientific understanding, and trust. Trust emerges as a central theme in watershed management, with notable differences between organizations participating in partnerships and those operating independently. Trust is measured along two dimensions – trust in social and non-governmental stakeholders, and trust in governmental stakeholders – highlighting how collaborative relationships shape perceptions and interactions within watershed governance. Organizations engaged in watershed partnerships consistently report higher trust across both dimensions, underscoring the positive impact of formal collaboration on mutual understanding. Federated organizations and regional NGOs are especially notable, often reporting normalized trust scores above 70 for social and nongovernmental stakeholders. One respondent attributes this increased trust to the improved awareness and communication that partnerships facilitate.

This pattern suggests that regular interactions and shared problem-solving within partnerships foster stronger stakeholder relationships. In contrast, trust dynamics among non-collaborating organizations are more complex and variable. While federated organizations often maintain relatively high trust in social stakeholders even without formal partnerships, other organization types show greater disparities. In particular, trust in governmental stakeholders tends to be significantly lower than trust in nongovernmental actors. This gap may result from fewer opportunities for direct engagement with government agencies, previous negative experiences with regulatory processes, or differing priorities regarding watershed management.

The absence of formalized communication channels and cooperative frameworks likely exacerbates these trust disparities. Trust patterns also vary with organizational capacity and scope of operations, reflecting differences in organizational priorities and stakeholder interactions. Entities with broader geographic reach often display different trust dynamics than locally focused organizations, reflecting their diverse experiences and challenges. One stakeholder highlights the importance of trust in overcoming resource limitations, remarking, "public trust can often bridge the gap where resources fall short." This observation underscores the critical role of trust as a component of social capital that compensates for material or operational constraints, particularly in resource-limited settings.

The interplay between collaboration and trust is mutually reinforcing, with partnerships playing a key role in enhancing both dimensions. Organizations within partnerships report higher levels of trust and a greater willingness to engage in joint initiatives and share resources. This dynamic creates what could be termed a "trust dividend," a virtuous cycle where increased trust fosters more effective collaboration, which in turn builds even greater trust (see Morris et al. 2013). However, respondents emphasized that trust-building is not automatic; it requires sustained effort, consistent communication, and a demonstrated commitment to shared goals. One participant notes that "trust relationships are maintained through regular communication and the follow-through on commitments," highlighting the effort required to nurture these relationships.

Several key factors emerged as instrumental in building and maintaining trust within watershed management contexts. Regular communication and transparent information sharing were cited as critical, alongside demonstrated follow-through on organizational commitments. Inclusive and transparent decision-making processes also played a vital role, fostering a sense of fairness and accountability among stakeholders. Recognizing interdependence among organizations and respecting differing perspectives and constraints further strengthened these trust relationships. Collectively, these factors provide a blueprint for building the trust necessary to achieve effective collaboration in watershed governance.

The implications of these trust dynamics for watershed management are significant. Higher trust levels among collaborating organizations create favorable conditions for information sharing, joint problem-solving, resource coordination, and long-term planning. Additionally, trust fosters effective conflict resolution, enabling organizations to navigate complex challenges more easily. The data also suggest that trust-building efforts should be prioritized in any strategy to improve watershed outcomes. These findings highlight the stark contrast in trust levels between collaborating and non-collaborating organizations, emphasizing the importance of formal partnerships for building the social capital necessary for effective governance. Partnerships provide a framework for trust development and create a foundation for sustainable and collaborative watershed management.

7.2.5 Scientific Understanding and Watershed Knowledge

The analysis of scientific understanding in watershed management reveals distinct differences between organizations engaged in collaborative partnerships and those operating independently. Scientific understanding is assessed through two primary dimensions: comprehension of watershed characteristics and resources, and understanding of problem causation within the watershed. These measures provide insights into how organizational collaboration influences the development and application of scientific knowledge in watershed management.

Organizations in collaborative partnerships demonstrate consistently higher levels of confidence in their scientific understanding than non-collaborating organizations. Federated organizations and regional NGOs report strong scientific knowledge, frequently scoring at or above 70 on the normalized scale. This elevated confidence stems from collaborative advantages, including shared research resources, regular information exchange, collective data interpretation, enhanced monitoring capabilities, and shared learning experiences. These factors contribute to individual and organizational capacity and an integrated understanding of watershed dynamics.

The relationship between collaboration and scientific understanding manifests across multiple operational areas. Partnership organizations report higher confidence in understanding watershed dynamics, identifying problem causes, interpreting environmental data, applying scientific findings to management decisions, and evaluating intervention effectiveness. These capabilities are essential to effective watershed management and demonstrate how collaborative approaches foster deeper scientific understanding.

By comparison, noncollaborating organizations report lower confidence in their scientific understanding, particularly regarding watershed problem causation. While federated organizations maintain relatively high confidence without formal partnerships, other groups like international NGOs show notably lower scores. This disparity indicates specific challenges faced by non-collaborating organizations, including limited scientific resource access, reduced knowledge-sharing opportunities, fewer data collection channels, isolated interpretation processes, and constrained monitoring capabilities.

The survey identifies varying confidence levels across different types of scientific knowledge. Organizations report the highest confidence in understanding basic watershed characteristics and moderate confidence in ecosystem interactions. However, confidence decreases significantly in areas requiring comprehension of complex causal relationships and long-term trends. One stakeholder emphasizes this challenge: "Monitoring seldom gets the funding, but [we] need it very much." This observation highlights the critical relationship between resource allocation and scientific knowledge development.

These findings have important implications for watershed management practice. Partnerships appear to significantly enhance organizations' capacity to develop and apply scientific understanding through comprehensive monitoring and analysis approaches. The disparity in scientific understanding between collaborating and non-collaborating organizations suggests potential benefits from expanding partnership opportunities to include more stakeholders. The results underscore that scientific understanding functions as both a foundation for and an outcome of effective collaboration in watershed management.

7.3 What Shapes Collaborative Engagement?

Experience with collaborative projects primarily influences how organizations view stakeholder involvement in watershed management. Results suggest that organizations with collaborative experience demonstrate that these past initiatives substantially inform their understanding of engagement requirements. Those with successful collaborations typically advocate for broader participation, having observed how diverse stakeholder involvement contributes to positive outcomes. Organizations with implementation challenges often emphasize specific needs for enhanced coordination and communication protocols, reflecting lessons learned through direct experience.

The geographic scope of operations also plays a fundamental role in shaping organizational perspectives on whether collaboration matters, and whether it makes a difference. Organizations operating across extensive territories consistently view watershed issues as fundamentally interconnected, leading them to estimate higher necessary levels of stakeholder involvement. Their broader operational scope enables the recognition of how actions in different watershed areas affect overall environmental conditions. In contrast, organizations focused on localized areas often perceive more contained challenges, resulting in lower participation estimates and more targeted engagement strategies.

An understanding of watershed dynamics is another critical factor influencing approaches to organizational stakeholder involvement. Organizations demonstrating comprehensive knowledge of hydrological systems typically advocate for broader participation, recognizing the complex interactions between ecological, social, and economic factors affecting watershed health. As one respondent notes, "it is crucial for people to have an emotional connection to the river, bay, or water body, for getting them involved in its protection or improvement." This observation reinforces that effective watershed management requires both technical expertise and strategies for meaningful community engagement (see Morris et al. 2013).

Organizational capacity significantly shapes how institutions approach stakeholder engagement in watershed management. Available staffing, funding, and technical capabilities directly influence how organizations conceptualize achievable levels of stakeholder involvement. Organizations with substantial resources often develop comprehensive engagement strategies incorporating diverse stakeholders across multiple sectors. Those with limited resources typically implement more targeted approaches focused on specific priorities or geographic areas, demonstrating how resource constraints can narrow the scope of engagement efforts.

Public engagement emerges as a crucial operational component, with clear distinctions between partner and non-partner approaches. Partnership organizations typically implement coordinated volunteer programs and large-scale community events, while non-partnering organizations focus on targeted educational initiatives and local outreach efforts. One stakeholder emphasizes this priority: "It is crucial for people to have an emotional connection to the river, bay, or water body for getting them involved in its protection or improvement." This observation highlights how effective watershed management requires technical expertise and sustained community participation to achieve environmental objectives.

7.3.1 Challenges to Collaboration Success

Imperial (2005, 281) cautions that watershed programs often exist in cases where governing responsibilities are scattered, problem-solving capacity is limited, and no single organization holds enough power to unilaterally solve "the problem." In fact, Imperial (2005) is best described as not sanguine about collaboration as a process to engender large-scale improvements in water quality, as it can be messy, slow-moving, and divisive, while its benefits are typically limited. That said, Imperial does acknowledge the promise of collaboration, especially for those involved (Ansell and Gash 2007; Bidwell and Ryan 2006; Koehler and Koontz 2008; Lubell et al. 2002). In these instances, perceptions matter. Believing the collaborative process is working can be key to sustaining it over time and to building the relationships and trust necessary for long-term change (Bardach 1998). We begin, however, with observations related to the challenges of collaboration.

Survey respondents estimate the percentage of stakeholders, for example, citizens, businesses, and governments, whose behavior or activities must change to resolve existing problems. This measure assesses how organizations conceptualize the scale of watershed issues and the degree of collective action necessary for effective solutions. Organizations involved in partnerships consistently provide higher estimates of required stakeholder behavioral change than non-collaborating organizations. These estimates typically exceed 60–70%, indicating that collaborative organizations recognize the extensive engagement needed to address watershed challenges. Partnership organizations demonstrate awareness of how addressing environmental issues requires coordinated action across multiple stakeholder groups.

Analysis of these estimations reveal several significant patterns. Regional NGOs and federated organizations within partnerships provide the highest estimates, emphasizing the need for changes across diverse stakeholder groups. Local government agencies and NGOs participating in partnerships report similarly high estimates, particularly in regions facing complex watershed problems. These organizations state that effective solutions require regulatory compliance and fundamental behavioral changes among community stakeholders.

Non-collaborating organizations report notably lower estimates of required stakeholder engagement, frequently below 50%. International NGOs and special districts demonstrate low estimates, suggesting limited integration with broader stakeholder networks. One respondent from a non-collaborating organization note, "the immediate problems seem to stem from a small group of stakeholders, but addressing root causes would require far more involvement than we can currently manage." This response reflects the constraints these organizations face in pursuing comprehensive solutions.

The consistent difference in estimates between collaborating and non-collaborating organizations highlights how partnerships deepen awareness of watershed management complexity. Through regular stakeholder engagement, shared learning, and coordinated action, organizations gain a broader perspective on the diverse ways stakeholders influence environmental outcomes. These results indicate that partnerships are instrumental in fostering more integrated and comprehensive watershed management strategies.

Specific Challenges Organizations face several significant operational challenges, as shown in Table 7.4. These exist alongside stakeholders' limited attention spans, full governing and organizational agendas, and fragmented decision-making. Fragmentation is particularly vexing in regulatory contexts, as one stakeholder observes: "Incremental and fragmented decision-making is inherent in the regulatory approach... It impedes political understanding and resolve."

Resource constraints represent another fundamental operational challenge, particularly affecting monitoring activities, as shown in Table 7.5. Despite the critical importance of monitoring to effective watershed management, organizations consistently report inadequate funding for these activities. As one participant notes, "Monitoring seldom gets the funding, but it needs it very much." This persistent resource limitation affects core activities from data collection to regulatory enforcement, creating a significant gap between management objectives and implementation capacity.

Watershed organizations across the country face a complex mix of natural and human-driven pressures that hinder effective management.

Table 7.4 *Examples of key challenges encountered by watershed collaborations, including climate impacts, stakeholder engagement, implementation costs, and public awareness.*

Collaboration	Examples
Friends of the Branch River	Increased storm events driven by climate change are exacerbated by weak regulatory standards and insufficient government monitoring of water quality degradation.
Adams County Land and Water Conservation Department	Difficulty securing participation from vegetable growers and dairy operations; overall participation in conservation efforts remains low.
Catoctin Land Trust	Collaboration is hampered by high implementation costs, lack of political will, and pressures from population growth and rising land values.
Ruby Habitat Foundation	Education remains a core challenge – many decision-makers who negatively impact the river may do so unknowingly. Greater awareness could lead to more protective behavior.

Source: Generated by authors.

Table 7.5 *Examples of resource-related challenges encountered by watershed collaborations, including infrastructure limitations, funding gaps, urban development, and community engagement issues.*

Collaboration	Examples of resource challenges
Project Watershed Central New York	Resource constraints include urban combined sewer overflows and agricultural runoff. A specific issue involves clay-heavy mud boils in a key tributary (Onondaga Creek). While sewage treatment facilities are currently adequate, sustained funding is uncertain. Chloride pollution from winter road deicing remains an unresolved challenge.
Alliance of Rouge Communities	Difficulty securing ongoing funding for issues considered "nobody's problem," such as habitat loss, sedimentation, and legacy pollution concerns.
Friends of the Rouge	Regional population growth has increased strain on sewage infrastructure, driven agricultural intensification, and accelerated conversion of forests and farmland to residential use – all placing stress on watershed health.
The River Raisin Institute	Limited funding hinders outreach and education efforts across the watershed. Apathy and lack of citizen engagement further exacerbate the difficulty of fostering stewardship and participation.

Source: Generated by authors.

Climate change is bringing more frequent and intense storms, droughts, and floods. These disruptions, together with new and emerging contaminants, are straining ecosystems and existing infrastructure. Collaboration is also hampered by governance challenges such as political resistance, bureaucratic inertia, community distrust, a tendency to prioritize short-term gains, and a preference for economic growth.

7.4 Conclusion

Understanding how decision levels interact reveals critical factors in real and perceived watershed management effectiveness. Constitutional arrangements, enable or constrain collective decision-making about resources and priorities, determining what operational activities are possible and how they are implemented. This multilevel perspective helps explain why similar operational activities may produce different outcomes across watersheds with different governance arrangements. Our framework thus provides essential insights for improving watershed management by highlighting how institutional structures and decision processes affect environmental results.

This analysis demonstrates the fundamental role of collaboration in watershed management effectiveness. The findings indicate that partnerships facilitate trust development, enhance scientific understanding, and expand community engagement capabilities. These collaborative mechanisms enable organizations to implement more comprehensive approaches to watershed governance through coordinated stakeholder action. Several key challenges emerge from this research that require attention to optimize collaborative frameworks. Fragmented governance structures, resource constraints, and complex stakeholder engagement demands necessitate systematic improvements in watershed management. Governance systems must evolve to better integrate across jurisdictions while incorporating diverse stakeholder perspectives. Stable funding mechanisms are essential to support critical activities like monitoring, and restoration, which underpin effective watershed management.

The research reveals important distinctions between partnering and non-partnering organizations in their operational approaches and effectiveness. Partner organizations demonstrate broader, more integrated perspectives on watershed challenges, reflecting the benefits of shared knowledge and resources. Non-partnering organizations typically operate within more constrained scopes, effectively addressing localized issues but showing limited capacity to engage with broader watershed complexities. This pattern suggests potential benefits from expanding partnership opportunities across watershed governance systems.

The findings contribute significant insights to watershed management practice while highlighting areas requiring further development. As environmental pressures on watershed systems increase, the need for coordinated and adaptive governance becomes more critical. Collaborative approaches demonstrate value in addressing complex watershed challenges through stakeholder coordination and resource sharing. However, these frameworks require careful attention to equity, transparency, and adaptability to maximize their effectiveness.

These conclusions suggest several important directions for future research and practice in watershed management. Future investigation should focus on refining collaborative strategies and expanding their applicability across diverse watershed contexts. Particular attention should be paid to developing governance mechanisms that facilitate cross-jurisdictional coordination while ensuring meaningful stakeholder participation. Such research can advance both theoretical understanding and practical implementation of effective watershed management approaches.

References

Ansell, Chris, and Gash, Alison. 2007. Collaborative Governance in Theory and Practice. *Journal of Public Administration Research and Theory*, **18**, 543–571.

Bardach, E. 1998. *Getting Agencies to Work Together: The Practice and Theory of Managerial Craftmanship*. Washington, D.C.: Brookings Institution.

Bidwell, Ryan D., and Ryan, Clare M. 2006. Collaborative Partnership Design: The Implications of Organizational Affiliation for Watershed Partnerships. *Society and Natural Resources*, **19**, 827–843.

Kash, Jeffrey P. 2008. Enemies to Allies: The Role of Policy-Design Adaptation in Facilitating a Farmer–Environmentalist Alliance. *Policy Studies Journal*, **36**(1), 39–60.

Koehler, Brandi, and Koontz, Thomas M. 2008. Citizen Participation in Collaborative Watershed Partnerships. *Environmental Management*, **41**, 143–154.

Lubell, Mark. 2003. Collaborative Institutions, Belief Systems, and Perceived Policy Effectiveness. *Political Research Quarterly*, **56**(3), 309–323.

Lubell, Mark. 2007. Familiarity Breeds Trust: Collective Action in a Policy Domain. *The Journal of Politics*, **69**(1), 237–250.

Morris, John C., Gibson, William A., Leavitt, William M., and Jones, Shana C. 2013. *The Case for Grassroots Collaboration: Social Capital and Ecosystem Restoration at the Local Level*. Lanham, MD: Lexington Books.

Ostrom, Elinor. 1990. *Governing the Commons: The Evolution of Institutions for Collective Action*. Cambridge, MA: Harvard University Press.

Ostrom, Elinor. 1998. A Behavioral Approach to the Rational Choice Theory of Collective Action. *American Political Science Review*, **92**(1), 1–22.

Ostrom, Elinor. 1999. Institutional Rational Choice: An Assessment of the Analysis and Development Framework. In: Sabatier, Paul A. (ed.), *Theories of the Policy Process*. Boulder, CO: Westview Press.

Ostrom, Elinor. 2005. *Understanding Institutional Diversity*. Princeton, NJ: Princeton University Press.

Sabatier, Paul A., Focht, Will, Lubell, Mark, Trachtenberg, Zev, Vedlitz, Arnold, and Matlock, Marty (eds.). 2005. *Swimming Upstream: Collaborative Approaches to Watershed Management*. Cambridge, MA: The MIT Press.

Scholz, John T., Berardo, Ramiro, and Kile, Brad. 2008. Do Networks Solve Collective Action Problems? Credibility, Search, and Collaboration. *The Journal of Politics*, **70**(2), 393–406.

Simon, Herbert. 1947. *Administrative Behavior: A Study of Decision-Making Processes in Administrative Organization*. New York: The Macmillan Company.

8

Turning Knowledge in Action

This book describes the physical and institutional landscape of watershed governance. Humans organize themselves around water – our habitats, our food production, our cities, our land distributions, and much of our energy production (Barham 2001). We are as much connected to our water resources as we are to our roads, homes, schools, and places of worship. As such, maintaining and sustaining water resources is as much an act of self-governance as it is an act of self-determination. Collaboration creates institutional arrangements that allow for the pursuit of numerous goals, create economies of scale, and distributes fairly the benefits and costs of collective action (Schlager and Blomquist 2008). As Tocqueville (1945) acknowledges, these institutional arrangements can be perceived as "wild," "checkered with accidents," but are often replete with "animation and effort." This description aligns with Environmental Protection Agency's (EPA) Watershed Approach, with its emphasis on improving watersheds "community by community, watershed by watershed" (EPA 1996) holds the potential for fostering institutionally rich environments with robust citizen engagement. The overarching question of this book, which looks at the potential of collaboration, asks whether this approach matters.

The urgency of addressing water resource management issues is clear. Increasing uncertainty due to climate change, intense weather events, droughts, floods, user demands, and nonpoint pollution are challenging the institutions designed to govern water quality. These more recent stressors exist alongside a fleet of challenges related to the realities of resource-constrained organizations and governments. Together, these challenges highlight the limitations inherent to top-down regulation and likely necessitate the deployment of alternative forms of environmental governance. One such option, and the focus of this book, is collaboration. Collaborative efforts are unique in that they transcend traditional governance boundaries and move stakeholders toward intersectoral partnerships and engagement to tackle twenty-first century water challenges.

The management of water quality in the twenty-first century involves disparate institutions, interests, politics, economics, and conflicts. The administration of water quality programs is further complicated because water policy involves an array of private property, economic development, food and agriculture, energy development, questions of environmental justice and protection, and community development interests. To distill it down, the topic of water is embedded in nearly everything and in every policy conversation. For some watersheds, and for some stakeholders, the response to these challenges is to collaborate so that they may deliver meaningful administrative outputs and achieve actual water quality improvements. But, as the analyses herein demonstrate, sometimes such efforts are not successful (at least in terms of revealed water quality outcomes) but are effective in contributing to a belief among stakeholders that water quality has improved.

Despite the growing calls from both academic and practitioner circles on the importance of collaborative approaches in environmental management there has, until this point, been a dearth of large-scale empirical studies that assess the revealed and perceived outcomes of watershed collaboration. To address this gap, this book presents the results of a mixed-methodological study that investigates measured water quality outcomes (to provide objective evidence of the impact of collaborative efforts) and perceived outcomes (to offer insights into stakeholders' attitudes and beliefs).

8.1 Watersheds and Collaboration

Each collaboration brings distinct and unique strengths, histories, experiences, weaknesses, and relationships to the challenge of protecting and/or restoring water quality. As noted, the federal government has established wide-ranging environmental policies, including cornerstone policies, which preclude a "race to the bottom" as well as promulgated enforcement mechanisms to ensure that quality thresholds are met. Voluntary frameworks exist alongside these policies. Whereas the federal government establishes the broad parameters of water quality, it is state and local agencies that are responsible for the day-to-day administration of water quality standards. At the subnational level, questions of authority, expertise, willingness, and resources are central and a necessary component to identifying, setting, and enforcing water quality uses and standards.

For point-source pollution, states and localities have historically managed resources, mobilized attention, prioritized certain bodies of water, and provided technical expertise (Elmore 1980). This approach has clearly resulted in improved water quality practices in business and industry, as well as a significant reduction in point-source pollution and traditional patterns of environmental discharges.

Yet, it has ultimately fallen short of meeting the aspirational goals identified in the Clean Water Act (CWA) as it relates to nonpoint source pollution. And, despite decades of work, water quality in the United States is at best uneven, suggesting that top-down approaches have struggled to implement solutions tailored to local conditions (Hupe 2011; Pressman and Wildavsky 1984). Watershed collaboration, by contrast, offers stakeholders with more direct access to decision-making processes as well as opportunities to build and maintain trust among participants. Within a collaborative framework, actors often possess a deep understanding of the local context and knowledge of watershed challenges, opportunities, threats, and strengths. Additionally, they are often tied to the region and/or the communities most directly impacted by the body of water in question (Fiorino 2001, 2006; Klyza and Sousa 2008; Morris et al. 2013).

Managing the nation's water quality, meeting the goals of the CWA, and addressing nonpoint source pollution are challenges that are never really solved. Ranging from agricultural runoff in the Midwest to mining pollution in the West and stormwater challenges in urban and suburban communities, water quality governance is a significant twenty-first century challenge. While scientific expertise and federal involvement are critical in identifying, prioritizing, and mitigating water pollution, equally important are the collaborative governance structures that shape outputs and outcomes. Within these structures are the classic questions of: Who is involved? Who should be involved? How are decisions made by those who govern? What factors contribute to successful outcomes? Despite the scope, scale, and cost of water-related challenges, a range of stakeholders are actively working to improve the nation's waters. Actors include state-level water administrators to rank-and-file members of place-based grassroots nonprofit groups, many of whom are building capacity and competencies, developing new areas of technical and social expertise, cultivating relationships with nonprofits and other stakeholders, and balancing competing goals and interests. Such efforts, as well as related successes, vary and the impacts of one program and/or effort often spills over to affect others (see Morris 2022).

8.2 What This Book Reveals

The use of collaboration (or lack thereof) has emerged and evolved alongside a more traditional mode of regulation. Whether the response is command-and-control regulation (see Ernst 2010) or collaboration (see Morris et al. 2013), water-related challenges cut across administrative and jurisdictional boundaries. Despite the promise of collaboration, as a "family" of interactions, the term enjoys a multitude of definitions, has roots in multiple disciplines, and often lacks conceptual clarity; for example, definitions of collaboration include up to 60 different elements (see Morris and Miller-Stevens 2016). Despite these shortcomings, the various strands of literature have coalesced around several themes or observations, pertinent here:

1. There is no definitive definition of collaboration and collaborative efforts vary widely in their origins, organizational structures, relationships with regulators and other actors, the type of participants and their duration, and decision-making processes. These differences, in turn, contribute to different organizational missions, strengths, and goals. As such, collaborations are likely to engender – or attempt to engender – different environmental and social outputs and outcomes as well as leverage a variety of strengths.
2. When collaboration does happen, the goals of collaborative efforts can vary substantially, as can the motivations for participation (Diaz-Kope and Morris 2019). While some groups may be focused on addressing watershed restoration, other groups may be more focused on watershed preservation or on a particular industry. Still others might be focused on the development of recreational opportunities. These different goals also make determinations of "success" somewhat problematic, in that "success" can only be measured in relation to the goals of the collaboration.
3. Collaboration is neither a panacea nor a substitute for top-down regulation. The risks and rewards of collaboration are best understood as fitting within a layered regulatory framework that stretches back two hundred plus years, and includes private property protections to classic environmental policies such as the Clean Water Act of 1972 and Water Quality Act of 1987. These eras and other policies hold numerous goals and priorities – some of which support collaboration, others impede collaboration, and others offer additional venues to effectuate changes in water quality. The point is that collaboration's outputs and outcomes are related to how the effort is situated within larger institutional, legal, and political systems.
4. Watershed collaboration faces several challenges. Threats include a range of emerging contaminants, the realities of aging infrastructure, and the vexatious nature of nonpoint source pollution and climate change. Each is compounding the difficulties of detecting, protecting, and/or treating water. Moreover, governance is characterized by numerous organizations, mismatched goals and boundaries, overlapping responsibilities, and multiple legal frameworks. Effective collaboration is further complicated by the presence of collective action problems, historical and ongoing conflict, and a lack of trust.

Water governance in the United States has emerged and evolved from a state enterprise to one that involves each level of government and many nongovernmental actors. Despite critical federal legislation including the Clean Water

Act of 1972, compliance challenges remain, even today. At the subnational level, enforcement is inconsistent, with many state and local governments struggling to meet the standards established by federal law. The emergence of collaborative watershed management resulted from several concurrent trends or realities: Watersheds being conceptualized as complex socioeconomic systems, the limitations inherent to traditional command-and-control regulation, the rise of climate change and nonpoint pollution, and the use of Section 303 of the CWA to compel state action.

Given this legal and political landscape, we then provide an overview of the literature on collaborative watershed governance. While definitions and elements abound, the literature generally treats watershed collaborations as a flexible and adaptive arrangement built on stakeholder knowledge, reciprocal trust and learning, and shared governance. We also elucidate key dimensions and characteristics associated with more effective forms of watershed collaboration, such as ecological conditions, social and civic characteristics, resources, and institutions. The latter part of the chapter then presents a model of watershed collaboration with an emphasis on water quality and social outputs and outcomes, both revealed and perceived. The central components include external factors, the role of information and information exchange, stakeholder characteristics, legitimacy, outputs or the tangible products or activities of collaboration, and revealed and perceived outcomes.

Subsequent chapters explore the relationship among watershed collaboration and revealed water quality outputs and outcomes across a sample of nearly 25% of all watersheds in the United States. Our model, described in Chapter 3, seeks to establish links between context, stakeholders, information communities, and legitimacy as a means to link actions to outputs and outcomes. These chapters rely on data from a large-n sample watersheds to analyze drivers of water quality improvement. Our findings point to several important factors in understanding watershed health. First, we find that many of the outputs of collaboration are statistically unrelated to revealed water quality improvements. That said, results point to important roles relative to historical water quality conditions and socioeconomic factors. Additionally, the presence of a regional nongovernmental organization is consistently associated with better results.

The book then shifts the focus toward stakeholder perceptions of water quality. Against the backdrop of limited water quality improvements, these chapters examine perceived outcomes via a survey of watershed organizations involved in our sample of watersheds. In general, three out of four respondents believe that watershed water quality improved and linked these improvements to the implementation of best management practices, increased engagement, and to other collaborative activities. Bivariate data also reveal that factors approximating the presence of local entities, watershed partnership, political efficacy, trust, and belief in the legitimacy of decision-making are correlated with positive perceptions. Sociodemographic characteristics are also associated with perceived watershed improvement. Yet, in more finely grained multivariate models, only watershed partnerships, population density, and racial composition remain statistically significant. These chapters also incorporate the voices of many of these stakeholders, often echoing the results of the survey. Respondents highlight how partnerships supported the development of more effective governance mechanisms, higher levels of trust, better and more comprehensive science, a more robust scientific understanding of various impacts, and more effective collaboration outputs and outcomes.

Overall, our analysis identifies three drivers of improved water quality outcomes in collaborative settings. First, regional-level nongovernmental organizations play a central role in improving revealed water quality outcomes. Second, history matters. In other words, historical water quality conditions have a large effect and influence on current water quality outcomes. Third, demographics may not be destiny, but they are important. Specifically, the percentage of the white population – significantly affects water quality, raising important considerations related to equity and historical investment and land use.

8.3 Does Collaboration Matter?

What is success when it comes to collaboration? How do we know success when we see it? These questions (and others like them) are at the center of watershed collaboration scholarship. The answers and perspectives likely depend on where one sits. From a scientific or ecological perspective, success is likely linked to improved water quality as measured through revealed pollutant reductions over a specific time period. Perceptions, however, may or may not match reality. By comparison, from a more social or governance perspective, success is likely to be intertwined with enhanced levels of social capital, participant engagement and learning as well as legitimacy. Yet watershed health may not be improved solely by changing social capital scores or participant learning.

We caution against the use of changes in measured water quality as the sole indicator of collaborative success. There are several reasons for this cautionary note. First, some water quality problems are intractable (or "wicked"; see Rittel and Webber 1973), and defy efforts to solve. In other words, water quality may deteriorate regardless of the efforts of a collaboration. It is also possible that, while the efforts may not result in improvements in water quality, the collaborative efforts may create stability in the watershed. To put it differently, while water quality may not improve, it may not get worse as a result of the collaborative efforts. Second, water quality data are notoriously incomplete, and problems abound in collection, analysis, and reporting (see Mayer et al. 2024; Aistrup et al. 2024). The cross-state variation in reporting frequency and completeness makes comparisons challenging. However, it is also worth noting that, in spite of the inherent limitations of these data, they remain the best available indicators of water quality across the fifty states.

Third, not all watershed collaborations are faced with the same environmental challenges, nor are their goals the same. In their study of three watershed collaborations in the Hampton Roads, Virginia region, Morris et al. (2013) documented both the environmental challenges in each watershed and the structure and goals of each collaborative group. The Elizabeth River was fraught with centuries of industrial pollution, and the goal of the Elizabeth River Project was to address segments of the waterway with targeted projects to address legacy pollution, such as the Money Point project. This was accomplished by incentivizing businesses and industries along the river to join the efforts. The goal was not to restore the river, but to reduce the effects of the legacy pollution. In comparison before the Lynnhaven River was the victim of twentieth-century suburban sprawl, and the nonpoint pollution resulting from lawn fertilizers, pet waste, leaking septic tanks, and sewer overflows. The goal was to be able to eat oysters from the river, which meant the implementation of best management practices (BMPs) through a public outreach and education model. Finally, the Nansemond River Preservation Alliance worked to protect the Nansemond River from degradation from encroaching suburban sprawl. The Nansemond was relatively clean, but was threatened by numerous housing developments along the river. In other words, different goals would lead to very different activities, and water quality measures would only be partial indicators of collaborative success.

To return to the book's central question: Does collaboration matter? The answer is yes, collaboration matters, but with caveats. Collaboration is not a simplistic, feel-good, or transformative mechanism or arrangement. Like more coercive forms of regulation, it is an arrangement that has clear strengths as well as obvious limitations. The evidence in this book, including stakeholder insights, survey findings, and revealed outcomes, suggest that collaboration plays a critical role, but one that depends on how success is defined. For social connections and perceptions of water quality outputs and outcomes, data is clear that collaboration matters a great deal. If success is defined as revealed water quality improvement, collaboration plays a more muted role and interacts with historical conditions, demographics, and institutional structures, although the presence of a regional organization certainly helps.

8.3.1 Collaboration as a Way Forward?

Our data are retrospective and show that collaboration matters. Yet will collaboration still matter as we move past the first quarter of the twenty-first century?

We surmise that it will continue to play a key role during the second Trump Administration and beyond, for reasons we outline in the following discussion.

First, today's water challenges are likely to remain or grow even more severe. Climate change, nonpoint source pollution, emerging contaminants, and aging infrastructure do not respect boundaries, thus making jurisdictionally focused approaches increasingly inadequate. Resources are likely to remain constrained and solutions dependent on multiple parties. In this environment, collaboration will be essential, as achieving (or maintaining) water quality will remain a complex challenge involving many interrelated dimensions, threats, governments, actors, and sectors.

Second, watershed collaboration is an alternative pathway available for state and local policymakers concerned with water quality. Such efforts operate within a multi-tiered legal and regulatory system that includes federal laws (e.g., the Clean Water Act 1972 Amendendments to the Federal Water Pollution Control Act (P.L. 92-500)) and state and substate laws, actors, and enforcement protocols. Often, collaboration emerges in policy areas with ambiguity, regulatory gaps, weak enforcement, and a need for coordination, resource mobilization, and legitimacy. These gaps, ambiguities, and supportive roles are likely to grow in the near and distant future, both in need and importance.

Third, collaboration generates a multitude of tangible products or outputs (watershed plans, MOUs) and outcomes (perceived or revealed improvements in water quality, increased trust). Outputs, however, do not guarantee desired outcomes. Yet such outputs are often prerequisites to achieving desired outcomes, as they build essential competences, momentum, and political efficacy that is necessary to sustain watershed health. By empowering subnational and substate actors, nonfederal actors are more likely to be empowered to take ownership of their local environment.

Fourth, as the federal government retreats from many of its Golden Age environmental commitments, collaboration may act as a buffer against this changing and uncertain environment. Recall that collaboration is often more agile and flexible as compared to top-down regulation. As such, collaboration enables stakeholders to adjust to an ever-changing legal, political, and environmental conditions by building on local knowledge, generating consensus, and realigning resources and goals in response to challenges.

Lastly, there are often strong perceived improvements in water quality when collaboration is present. Stakeholders linked these gains to best management practices, education, trust, and more robust governance mechanisms. These perceptions matter: They influence political will, resource flows, and sustained engagement, all characteristics that will continue to be important in the twenty-first century.

Collaboration matters. It is not a cure-all, but a vital complement to traditional regulation. Going forward, collaboration will likely be needed as the policymaking environment governing water quality is increasingly complex, conflictual, and fragmented. While our data suggest that the relationship between collaboration and water quality improvement is slight (but positive), the presence of collaboration leads to real and perceived social benefits. In short, collaboration may matter most by engendering trust, engaging stakeholders, building capacity, incorporating local knowledge, and building organizations that can adapt, all of which make protecting, preventing, and restoring a watershed possible.

8.4 Lessons Learned for Practice

Given the complexity of securing water quality, let alone water quality in an age of climate change, the path forward will demand leveraging the strengths of multiple levels of government and sectors (Ostrom 2009, 2010). The following sections explore the implications of our results to identify governance strategies that promote collaborative watershed governance. Such approaches build upon the principles of smart management, which encourage reflective decision-making, ongoing learning, and flexibility in response to changing conditions (Konschnik and Boling 2014).

8.4.1 Strengthening Capacity

Kettl (2006, 2016) argues that effective governance often starts with effective partnerships that include opportunities for participants to engage with and learn from one another. Effective capacity building involves supporting efforts to nurture open communication, collective problem-solving, and trust-building among stakeholders. By doing so, policymakers improve the likelihood that decisions reflect best available scientific knowledge and local context. Currently, however, many watershed collaborations – as well as collaborators – are episodic and decentralized. This dynamic may mean that collaborations lack the institutional structures needed to realize long-term water quality improvement. Strengthening collaborative capacity may include:

- Ongoing and open communication among state and local water managers with applicable nongovernmental organizations to build capacity and ensure alignment among multiple policy goals. This may involve the identification and sharing of scientific baselines that are necessary for water quality assessment.
- Promoting transparency and accountability through public access to applicable data.

- The development and/or continued support of region-wide watershed organizations or councils or other collaborative groups that bring stakeholders together to facilitate information sharing, goal setting, and administrative activities that support decision-making.
- The creation and use of technical as well as collaboration/relationship-building assistance programs to support stakeholders involved in the implementation of water quality programs. These can include the deployment of data-sharing efforts and other forms of information or the clarification of priorities to address specific concerns such as contamination, legacy pollution, erosion, or conservation.
- Resources that are designed to enhance trust, transparency, and adaptive management across stakeholders, especially in watersheds with competing goals related to economic, energy, environmental, and community interests.

8.4.2 Emphasizing Regional Watershed Approaches, Thinking, and Actors

Watersheds do not adhere to city, county, state, or national level boundaries. Thus, one option is the greater utilization of regional approaches that incorporate a variety of perspectives, especially as the presence of a regional association was associated with an increase in actual water quality. As noted previously, regional associations are linked with reductions in administrative obstacles, and conflicts, are associated with increases in efficiencies, and greater representation of various stakeholders. As applied to watersheds, regional efforts may be useful for planning and implementation related to:

- Navigating agreements with various users such as urban and agricultural users.
- Serving as a regional collaborative convener (McNamara 2016; McNamara and Morris 2012) to bring stakeholders to the table, and to develop pathways to successful collaborative processes.
- Supporting infrastructure projects (e.g., stormwater retention basins).
- Coordinating water quality monitoring and data collection programs to track contamination levels.
- Developing nontraditional partnerships with public, private, and non-profit sector organizations to enhance capabilities (Leavitt and Morris 2007).

8.4.3 Adaptive Agreements

Clarity in watershed collaboration is crucial for avoiding conflicts and disputes. Overly rigid regulations can depress participation, increase costs, and make compliance more difficult. Yet, because water is subject to multiple uses, the presence of vague or ambiguous language can be problematic and contribute to litigation (Matland 1995). One strategy for enhancing policy clarity is through adaptive agreements and other forms of technology that encourage stakeholders to be nimble. Possibilities include:

- Memoranda of Understanding (MOUs) that establish clear directives and thresholds for watershed management and administrative actions.
- Performance-based regulatory frameworks that allow flexibility in achieving water quality goals.
- Tiered incentive schemes that encourage voluntary conservation practices.

8.4.4 Leveraging Technology

The terrain of collaborative watershed governance is beginning to intersect with new forms of technology, most notably blockchain technology. At first glance, blockchain's reputation for hype may seem out of place in the slow, locally embedded work of watershed partnerships. Yet the underlying problems that blockchain aims to solve – persistent information gaps, fragmented records, and limited trust across organizational boundaries – are deeply familiar to anyone engaged in multi-party collaboration.

Blockchain's basic premise is simple: It is a distributed ledger system that allows each participant to record, verify, and audit transactions or data in a way that is transparent and extremely difficult to alter after the fact. In the context of watershed management, this could mean maintaining an open, shared record of grant allocations, BMPs adoption, or water quality monitoring results, reducing opportunities for confusion or dispute. The technology is less about novelty for its own sake, and more about building trust and reliability into the day-to-day workings of collaboration, particularly in environments where authority and responsibility are diffuse and past grievances may linger.

Still, blockchain is no silver bullet. Its value depends entirely on how it is integrated into the messy realities of collaborative work. Poorly designed, it risks hardening boundaries, excluding stakeholders who lack digital access or expertise, or locking in rules that cannot adapt to changing needs (Adams 2025). It cannot conjure consensus where interests fundamentally diverge, nor can it replace the face-to-face negotiation and iterative learning on which genuine collaboration depends. As with any tool, its effects will be shaped by existing institutional dynamics and the willingness of partners to share power and information.

If used thoughtfully, however, blockchain may serve as a new layer of infrastructure, helping to reinforce transparency, mutual accountability, and shared authority among actors who may never fully trust one another, but who must nevertheless work together. The broader lesson, echoed throughout this book, is that tools and technology are only as effective as the social and institutional foundations on which they rest. Blockchain's promise is not to eliminate the challenges of collaboration, but to make joint action more visible, auditable, and – at best – more resilient in the face of uncertainty.

8.4.5 Doubling Down on Learning

Due to the incremental nature of policy making, as well as the lack of a comprehensive national approach to water governance, the policies and programs governing watersheds are highly fragmented at the federal, state, and often local levels. This institutional fragmentation "limits any organization's ability to accomplish its mission by acting alone" (Imperial 2005, 283). Much of the collaborative activity taking place in watersheds today involves the education of resource users and programs to mitigate the spread of nonpoint source (NPS) pollution into streams, rivers, lakes, and estuaries – this, we suggest, must be continued. In other words, because of the heterogeneous nature of NPS pollution its diffuse sources, collaboration, as well as efforts to learn and engage, are often required. Moreover, due to the nature of the federal system in the United States, as well as to the diversity of land ownership and property rights, these efforts often require actors from multiple levels of government and the private sector to come together and learn from one another.

8.4.6 Focus on Finances

In many watersheds, past contamination still challenges current uses and users. As noted previously, one of the largest obstacles to achieving water quality is the lack of consistent and adequate funding and/or the capacity to secure funding. Many watershed collaborations rely on volunteers, short-term or temporary funding which, while valuable, may contribute to uncertainty, turnover, and a limited ability to restore contaminated waters and/or protect pristine bodies of water. To mitigate legacy pollution, watershed collaborations may consider diverse and secure funding mechanisms as well as partnerships with applicable agencies, as follows:

- Grant programs, such as the EPA Clean Water Act Section 319 grants, U.S. Department of Agriculture (USDA) conservation programs, and state-level watershed restoration funds. Other funding sources may include larger nonprofits and foundations whose missions align with watershed restoration and protection.
- Public-Private Partnerships (PPPs) that enable collaborations to cofund watershed protection efforts.

- Watershed Funds (or supporting such funds) that generate revenues via usage or impact fees or tax dollars to provide more stable funding.

8.5 Future Research Directions

Based on this research, we can identify several different avenues for future research. While much is known about collaboration, somewhat less is known about what makes collaboration successful or not. In a comprehensive review of the literature, McNamara, Miller-Stevens, and Morris (2020) capture the elements of what made collaboration "successful." They report a large number of factors gleaned from the literature, each proclaiming to be the "key" to collaborative success. On the other hand, very little appears in the literature to examine why collaboration fails. The takeaway from this is that there is little agreement in the literature about what constitutes success or failure; we would also suggest that one of the reasons for this state is that our theoretical understanding of collaboration is still in its infancy. Any study (including this study) that speaks to collaborative success or failure includes some set of underlying assumptions about how to conceive of success or failure in this setting. Our findings illustrate the challenges inherent in this issue.

A more nuanced study could also attempt to differentiate between collaborative goals. While this would be a very useful activity, it is fraught with difficulty. First, collaborative groups may have more than one goal. Is there a goal that takes precedence over others? Are the goals mutually supportive, or are they mutually exclusive? Are the goals reasonable, or measurable? How does the group define "success?" The researcher can impose a goal (as we have done here), but such an approach is necessarily imprecise. However, it is one of the tradeoffs of a large-*n* study. In order to conduct such a study, it is necessary to make some simplifying assumptions about activity in the setting under study. To the extent goals can be articulated by the individual collaboratives, perhaps subanalyses could look for characteristics that lead to greater success rates in the goals as articulated by the participants.

As noted several times in this book, a significant challenge to work in this arena is the inconsistent nature of water quality data. The percentage of state waterways assessed varies significantly, as does the regularity of the measurement of water quality in those waters. Likewise, states are not always vigilant about reporting their water quality data to EPA, making both cross-sectional and time-series studies problematic because of missing data. Most states do not measure their waterways every year, but measure segments on a rotating basis, making year-to-year comparisons challenging. Our review of the Assessment Total Maximum Daily Load (TMDL) Tracking and Implementation System (ATTAINS) maintained by EPA (2025) illustrates this problem. Some states with salt-water coastlines report having no coastal waters; other states report significant changes in the miles of rivers or miles of coastline in their state from year to year. The data quality problems make state-to-state comparison challenging (see Mayer et al. 2024). The good news is that focusing on specific segments of a waterway over time is somewhat less problematic from a data perspective, but there is still the problem of inconsistent measurement and reporting over time. Future research could seek to identify alternative (but comparable) sources of data that could be tracked over time to measure changes in water quality. Data might exist in state environmental or health agencies, for example. The data collection for such a project would be significant work, but might provide a more consistent picture of changes in measured water quality.

A promising avenue for future research involves integrating remote sensing data to visualize environmental change in collaborative watersheds. Specifically, satellite imagery (such as Landsat-derived Normalized Difference Vegetation Index (NDVI)) can track changes in riparian "greenness" over time, serving as a proxy for vegetation restoration, land-use shifts, or ecological health in response to collaborative interventions. This method enables researchers to capture spatially explicit environmental change, particularly in areas where traditional water quality metrics may be unavailable or insufficient. When paired with collaborative activity data, NDVI time series can offer insights into the ecological footprint of institutional arrangements. Prior research has demonstrated the utility of NDVI in tracking riparian restoration and land management outcomes (Pettorelli et al. 2005), making it a compelling tool for longitudinal analysis of watershed governance impacts.

Another arena ripe for future research is to test the feedback loop present in our theoretical model, depicted in Figure 3.2. While our study does test links between collaborative action and outputs and outcomes, it is not clear how the changes (or lack of changes) in the watershed impact the context at a later point in time (t+1). We speculate that watershed changes lead to a revised context, which in turn affects future collaborative efforts. This is a common element of models of watershed collaboration (see, e.g., Morris, et al. 2013; Sabatier, et al. 2005), but it is an element that is largely untested in the literature. To test this element would require a longitudinal study design with multiple points of measurement, but such a study would provide important evidence about the efficacy of watershed collaboration over time.

Finally, using water quality as the sole measure of "success" of watershed collaboration provides an incomplete picture of the benefits of collaboration. Numerous studies have pointed to the benefits of collaboration, and its ability to strengthen community ties, increase social capital, and bring organizations from different sectors of society together for a common purpose. We are well aware of the challenges to define such benefits and to collect meaningful (and comparable) data to capture these elements, but the benefits of doing so are undeniable. There are numerous case studies that address these issues, but we are not aware of any large-*n* studies that attempt to address these elements of watershed collaboration.

8.6 Concluding Thoughts

This research begins the process of filling a critical gap in the existing environmental management, collaboration, and policy literatures. It does so by offering a comprehensive and rigorous examination of the role and impact of collaboration in watershed management outcomes across the United States. Its unique contribution lies in its mixed methodology including a large-*n* study approach, which provides a more representative understanding of the dynamics and outcomes of collaborative watershed management for nearly one-quarter of all US watersheds. In short, we examine the relationship between watershed stakeholders and water quality outcomes (both revealed and perceived) in a sample that represents nearly 25% of all watersheds across the United States.

We began with the goal to better understand the dynamics of collaboration for environmental and natural resource management. Is this approach worth it? Do collaborative outputs produce beneficial outcomes? These are questions worthy of investigation, and their answers provide a genuine contribution to the public administration and public policy literatures and to practitioners. Our answer is – yes, collaboration matters. Of course, further exploration is needed to determine if we are to if and how these institutional arrangements fulfill their lofty goals. This is of more practical concern than theoretical. From water scarcity in the Western US and around the globe, to water quality concerns due to increased consumption of energy, food, and industrial goods, to aging infrastructure delivering water to homes, businesses, school, and hospitals, our water resources are strained. Collaborative approaches offer the promise to expand these constraints and improve these conditions, both social and ecological.

References

Adams, David P. 2025. The Code Is the Policy: Blockchain Governance and the Authoritarian Pursuit of Administrative Efficiency. *Administrative Theory & Praxis*, November, 1–11. doi:10.1080/10841806.2025.2585729.

Aistrup, Joseph A., and Morris, John C. 2024. Does Institutional Location Matter? The Effect of Agency Type on Resource Distribution for Nonpoint Water Pollution. *Water Policy*, **26**(10), 1039–1055.

Barham, Elizabeth. 2001. Ecological Boundaries as Community Boundaries. *Society and Natural Resources*, **14**, 181–191.

Diaz-Kope, Luisa, and Morris, John C. 2019. *Organizational Motivation for Collaboration: Theory and Evidence*. Lanham, MD: Lexington Books.

Environmental Protection Agency. 1996 (June). *Watershed Approach Framework*. Tech. rept. United States Environmental Protection Agency.

Environmental Protection Agency. 2025. *ATTAINS*.

Ernst, Howard. 2010. *Fight for the Bay: Why a Dark Green Environmental Awakening is Needed to Save the Chesapeake Bay*. Lanham, MD: Rowman and Littlefield.

Fiorino, Daniel J. 2001. Environmental Policy as Learning: A New View of an Old Landscape. *Public Administration Review*, **61**(3), 322–334.

Fiorino, Daniel J. 2006. *The New Environmental Regulation*. MIT Press.

Hupe, P. L. 2011. The Thesis of Incongruent Implementation: Revisiting Pressman and Wildavsky. *Public Policy and Administration*, **26**(1), 63–80.

Kettl, Donald F. 2006. Managing Boundaries in American Administration: The Collaboration Imperative. *Public Administration Review*, **66**, 10–19.

Kettl, Donald F. 2016. *Politics of the Administrative Process*. Washington D.C.: CQ Press.

Klyza, C. M., and Sousa, D. 2008. *American Environmental Policy, 1990–2006: Beyond Gridlock*. Cambridge: MIT Press.

Konschnik, K. E., and Boling, M. K. 2014. Shale Gas Development: A Smart Regulation Framework. *Environmental Science & Technology*, **48**(15), 8404–8416.

Leavitt, William M., and Morris, John C. 2007. Public Works Service Arrangements in the 21st Century: The Multiple-Sector Partnership as an Alternative to Privatization. *Public Works Management & Policy*, **12**(1), 325–330.

Matland, Richard E. 1995. Synthesizing the Implementation Literature: The Ambiguity-Conflict Model of Policy Implementation. *Journal of Public Administration Research and Theory*, **5**(2), 145–174.

Mayer, Martin K., Morris, John C., Zhang, Xiaodan, and McNamara, Madeleine W. 2024. Explaining State Efforts to Create TMDL (Total Maximum Daily Load) Agreements. *Social Science Quarterly*, **105**(5), 1776–1790.

McNamara, Madeleine. 2016. Collaborative management leadership: A skill set for the Entrepreneur. In *Advancing Collaboration Theory: Models, Typologies, and Evidence*, edited by John C. Morris and Katrina Miller-Stevens, 116–132. New York: Routledge.

McNamara, Madeleine W., Katrina Miller-Stevens, and John C. Morris. 2020. Exploring the Determinants of Collaboration Failure. *International Journal of Public Administration*, **43**(1), 49–59.

McNamara, Madeleine, and Morris, John C. 2012. More than a “one-trick pony”: Exploring the contours of a multi-sector convener. *Journal of Nonprofit Management*, **15**(1), 84–103.

Morris, John C. 2022. *Clean Water Policy and State Choice: Promise and Performance in the Water Quality Act*. Cambridge University Press.

Morris, John C., and Miller-Stevens, Katrina. 2016. The State of Knowledge in Collaboration. In *Advancing Collaboration Theory: Concepts and Cases*, edited by John C. Morris and Katrina Miller-Stevens, 3–13. New York: Routledge.

Morris, John C., Gibson, William A., Leavitt, William M., and Jones, Shana C. 2013. *The Case for Grassroots Collaboration: Social Capital and Ecosystem Restoration at the Local Level*. Lanham, MD: Lexington Books.

Ostrom, Elinor. 2009. A General Framework for Analyzing Sustainability of Social-Ecological Systems. *Science*, **325**(5939), 419–422.

Ostrom, Elinor. 2010. Beyond Markets and States: Polycentric Governance of Complex Economic Systems. *American Economic Review*, **100**(3), 641–672.

Pettorelli, Nathalie, Vik, Jon O.; Mysterud, Atle, Gaillard, Jane M., Tucker, Cédric J., and Stenseth, Nils Chr. 2005. Using the Satellite-Derived NDVI to Assess Ecological Responses to Environmental Change. *Trends in Ecology & Evolution*, **20**(9), 503–510.

Pressman, J. L., and Wildavsky, A. 1984. *Implementation: How Great Expectations in Washington Are Dashed in Oakland; Or, Why It's Amazing that Federal Programs Work at All, This Being a Saga of the Economic Development Administration as Told by Two Sympathetic Observers Who Seek to Build Morals on a Foundation*. Vol. 708. University of California Press.

Schlager, Edella, and Blomquist, William. 2008. *Embracing Watershed Politics*. Boulder, CO: University Press of Colorado.

Tocqueville, Alexis de. 1945. *Democracy in America*. New York: Alfred A. Knopf. Originally published 1835–1840.

Appendix

Watershed Stakeholder Survey

Participation

Thank you for your interest in this questionnaire. Your responses will provide valuable insight into the management and governance of our nation's water resources.

Do you wish to participate in this survey?

- Yes
- No

Organization Name and Type

What is the name of your organization?

[Text field]

Which of the following best describes your organization?

- Local Governmental Agency
- Special Government District (e.g., water district, conservation district)
- State Governmental Agency
- National Governmental Agency
- Tribal Government / Agency
- Regional Planning Agency
- Nongovernmental Organization (e.g., Natural Resources Defense Council)
- Nonprofit (e.g., 501(c)(3))
- Private Association (e.g., Friends of Lake Arrowhead)
- Other

Approximately how many full-time staff are employed by your organization?

[Text field]

Approximately how many part-time staff are employed by your organization?

[Text field]

Approximately how many people volunteer for your organization in an average year?

[Text field]

Permits and Licenses

Does your organization issue permits to resource users?

- Yes
- No

Does your organization issue licenses to resource users?

- Yes
- No

HUCs (Hydrologic Unit Codes)

For this study, watersheds are conceptualized at the HUC8, or cataloging unit, level.

In the following questions, please select your region and the watershed(s) in which your organization participates.

If your organization is involved in multiple watersheds, you will have the opportunity to select a single watershed to discuss for the purposes of this survey.

In which region does your organization work? [Dropdown menu]

Watershed Selection by Region

New England Region (01)

- 01010004 – Aroostook
- 01020003 – Mattawamkeag
- 01030002 – Dead
- 01050001 – St. Croix
- 01060001 – Presumpscot
- 01070002 – Winnipesaukee River
- 01070006 – Merrimack River
- 01080104 – Upper Connecticut-Mascoma
- 01080201 – Middle Connecticut
- 01080205 – Lower Connecticut
- 01090001 – Charles
- 01100001 – Quinebaug

- 01100005 – Housatonic
- Other

Mid Atlantic Region (02)

- 02020003 – Hudson-Hoosic
- 02020007 – Rondout
- 02030103 – Hackensack-Passaic
- 02030202 – Southern Long Island
- 02040103 – Lackawaxen
- 02040201 – Crosswicks-Neshaminy
- 02040205 – Brandywine-Christina
- 02040302 – Great Egg Harbor
- 02050102 – Chenango
- 02050106 – Upper Susquehanna-Tunkhannock
- 02050203 – Middle West Branch Susquehanna
- 02050301 – Lower Susquehanna-Penns
- 02050305 – Lower Susquehanna-Swatara
- 02060003 – Gunpowder-Patapsco
- 02070001 – South Branch Potomac
- 02070005 – South Fork Shenandoah
- 02070009 – Monocacy
- 02080102 – Great Wicomico-Piankatank
- 02080106 – Pamunkey
- 02080110 – Tangier
- 02080203 – Middle James-Buffalo
- 02080207 – Appomattox
- Other

South Atlantic-Gulf Region (03)

- 03010103 – Upper Dan
- 03010107 – Lower Roanoke
- 03010204 – Meherrin
- 03020103 – Lower Tar
- 03020202 – Middle Neuse
- 03020302 - New River
- 03030005 – Lower Cape Fear
- 03040102 – South Yadkin
- 03040201 – Lower Pee Dee
- 03040205 – Black
- 03050101 – Upper Catawba

- 03050105 – Upper Broad
- 03050109 – Saluda
- 03050201 – Cooper
- 03050206 – Four Hole Swamp
- 03050209 – Bulls Bay
- 03060103 – Upper Savannah
- 03060107 – Stevens
- 03060201 – Upper Ogeechee
- 03070101 – Upper Oconee
- 03070105 – Little Ocmulgee
- 03070202 – Little Satilla
- 03080101 – Upper St. Johns
- 03080202 – Cape Canaveral
- 03090103 – Western Okeechobee Inflow
- 03090204 – Big Cypress Swamp
- 03100102 – Myakka
- 03100203 – Little Manatee
- 03100207 – Crystal-Pithlachascotee
- 03110103 – Aucilla
- 03110204 – Little
- 03120002 – Upper Ochlockonee
- 03130003 – Middle Chattahoochee-Walter F. George
- 03130007 – Kinchafoonee-Muckalee
- 03130011 – Apalachicola
- 03140101 – St. Andrews-St. Joseph Bays
- 03140105 – Pensacola Bay
- 03140202 – Pea
- 03140303 – Sepulga
- 03150102 – Coosawattee
- 03150106 – Middle Coosa
- 03150110 – Lower Tallapoosa
- 03150204 – Lower Alabama
- 03160104 – Tibbee
- 03160108 – Noxubee
- 03160112 – Upper Black Warrior
- 03160203 – Lower Tombigbee
- 03170002 – Upper Chickasawhay
- 03170006 – Pascagoula
- 03180001 – Upper Pearl
- 03180005 – Bogue Chitto

- Other

Great Lakes Region (04)

- 04010202 – Cloquet
- 04020102 – Ontonagon
- 04020201 – Betsy-Chocolay
- 04030101 – Manitowoc-Sheboygan
- 04030105 – Peshtigo
- 04030109 – Cedar-Ford
- 04030201 – Upper Fox
- 04040001 – Little Calumet-Galien
- 04050002 – Black-Macatawa
- 04050006 – Lower Grand
- 04060103 – Manistee
- 04060107 – Brevoort-Millecoquins
- 04070004 – Cheboygan
- 04080101 – Au Gres-Rifle
- 04080201 – Tittabawassee
- 04080205 – Cass
- 04090003 – Clinton
- 04100002 – Rasin
- 04100006 – Tiffin
- 04100010 – Cedar-Portage
- 04110002 – Cuyahoga
- 04120102 – Cattaraugus
- 04130002 – Upper Genesee
- 04140201 – Seneca
- 04150102 – Chaumont-Perch
- 04150304 – Grass
- 04150308 – Chateaugay-English
- 04150404 – Ausable River
- 04150408 – Lake Champlain
- Other

Ohio Region (05)

- 05010002 – Conewango
- 05010006 – Middle Allegheny-Redbank
- 05020001 – Tygart Valley
- 05020005 – Lower Monongahela
- 05030103 – Mahoning

- 05030201 – Little Muskingum-Middle Island
- 05040001 – Tuscarawas
- 05040005 – Wills
- 05050003 – Greenbrier
- 05050007 – Elk
- 05060002 – Lower Scioto
- 05070201 – Tug
- 05080001 – Upper Great Miami
- 05090102 – Twelvepole
- 05090202 – Little Miami
- 05100201 – North Fork Kentucky
- 05100205 – Lower Kentucky
- 05110004 – Rough
- 05120102 – Salamonie
- 05120106 – Tippecanoe
- 05120110 – Sugar
- 05120114 – Little Wabash
- 05120203 – Eel
- 05120207 – Muscatatuck
- 05130102 – Rockcastle
- 05130106 – Upper Cumberland-Cordell Hull Reservoir
- 05130202 – Lower Cumberland-Sycamore
- 05130206 – Red
- 05140104 – Blue-Sinking
- 05140204 – Saline
- Other

Tennessee Region (06)

- 06010102 – South Fork Holston
- 06010106 – Pigeon
- 06010202 – Upper Little Tennessee
- 06010206 – Powell
- 06020002 – Hiwassee
- 06030002 – Wheeler Lake
- 06030006 – Bear
- 06040004 – Buffalo
- Other

Upper Mississippi Region (07)

- 07010102 – Leech Lake

- 07010106 – Crow Wing
- 07010202 – Sauk
- 07010206 – Twin Cities
- 07020003 – Lac Qui Parle
- 07020007 – Middle Minnesota
- 07020011 – Le Sueur
- 07030003 – Kettle
- 07040002 – Cannon
- 07040006 – La Crosse-Pine
- 07050002 – Flambeau
- 07050006 – Eau Claire
- 07060003 – Grant-Little Maquoketa
- 07070001 – Upper Wisconsin
- 07070005 – Lower Wisconsin
- 07080103 – Lower Wapsipinicon
- 07080107 – Skunk
- 07080204 – West Fork Cedar
- 07080208 – Middle Iowa
- 07090003 – Pecatonica
- 07090007 – Green
- 07100004 – Middle Des Moines
- 07100008 – Lake Red Rock
- 07110003 – South Fabius
- 07110007 – Salt
- 07120002 – Iroquois
- 07120006 – Upper Fox
- 07130003 – Lower Illinois-Lake Chautauqua
- 07130007 – South Fork Sangamon
- 07130011 – Lower Illinois
- 07140103 – Bourbeuse
- 07140107 – Whitewater
- 07140203 – Shoal
- Other

Lower Mississippi Region (08)

- 08010202 – Obion
- 08010206 – Forked Deer
- 08010210 – Wolf
- 08020202 – Upper St. Francis
- 08020301 – Lower White-Bayou Des Arc

- 08020401 – Lower Arkansas
- 08030202 – Tallahatchie
- 08040101 – Ouachita Headwaters
- 08040206 – Bayou D'Arbonne
- 08040303 – Dugdemona
- 08050001 – Boeuf
- 08060201 – Upper Big Black
- 08060205 – Homochitto
- 08070202 – Amite
- 080703000 – Lower Grand
- 08080201 – Mermentau Headwaters
- 08080205 – West Fork Calcasieu
- 08090202 – Lake Pontchartrain
- Other

Souris-Red-Rainy Region (09)

- 09010002 – Des Lacs
- 09010006 – Long Creek
- 09020102 – Mustinka
- 09020106 – Buffalo
- 09020201 – Devil's Lake
- 09020205 – Maple
- 09020304 – Thief
- 09020308 – Forest
- 09020312 – Two Rivers
- 09030001 – Rainy Headwaters
- 09030006 – Big Fork
- 09040001 – St. Mary's
- Other

Missouri Region (10)

- 10020003 – Ruby
- 10020007 – Madison
- 10030103 – Smith
- 10030202 – Cut Bank
- 10040101 – Bullwhacker-Dog
- 10040105 – Big Dry
- 10040203 – Flatwillow
- 10050002 – Upper Milk
- 10050006 – Sage

- 10050010 – Cottonwood
- 10050014 – Beaver (Milk Basin)
- 10060002 – Redwater
- 10060006 – Big Muddy
- 10070003 – Shields
- 10070007 – Upper Yellowstone-Pompey's Pillar
- 10080003 – Popo Agie
- 10080007 – Upper Bighorn
- 10080011 – Dry
- 10080015 – Lower Bighorn
- 10090201 – Middle Fork Powder
- 10090205 – Crazy Woman
- 10090209 – Lower Powder
- 10100003 – Rosebud
- 1010102 – Little Muddy
- 10110204 – Beaver (Little Missouri Basin)
- 10120103 – Upper Cheyenne
- 10120107 – Beaver (Cheyenne Basin)
- 10120111 – Middle Cheyenne-Elk
- 10120202 – Lower Belle Fourche
- 10130103 – Apple
- 10130201 – Knife
- 10130205 – Cedar
- 10130303 – Grand
- 10140101 – Fort Randall Reservoir
- 10140105 – Crow
- 10140204 – Lower White
- 10150004 – Middle Niobrara
- 10160001 – James Headwaters
- 10160005 – Mud
- 10160009 – Turtle
- 10170103 – Lake Thompson
- 10170204 – Rock
- 10180004 – Medicine Bow
- 10180008 – Glendo Reservoir
- 10180012 – Horse
- 10190002 – Upper South Platte
- 10190006 – Big Thompson
- 10190010 – Kiowa
- 10190014 – Pawnee

- 10190018 – Lower South Platte
- 10200201 – Lower Platte-Shell
- 10210002 – Dismal
- 10210006 – Upper North Loup
- 10210010 – Cedar
- 10220004 – Logan
- 10230004 – Monona-Harrison Ditch
- 10240001 – Keg-Weeping Water
- 10240005 – Tarkio-Wolf
- 10240009 – West Nodaway
- 10240013 – One Hundred and Two
- 10250004 – Upper Republican
- 10250008 – Medicine
- 10250012 – South Fork Beaver
- 102500016 – Middle Republican
- 10260003 – Upper Smoky Hill
- 10260007 – Big
- 10260011 – Upper North Fork Solomon
- 10260015 – Solomon
- 10270104 – Lower Kansas
- 10270204 – Turkey
- 10280101 – Upper Grand
- 10280202 – Lower Chariton
- 10290103 – Little Osage
- 10290107 – Pomme De Terre
- 10290111 – Lower Osage
- 10300101 – Lower Missouri-Crooked
- 10300200 – Lower Missouri
- Other

Arkansas-White-Red Region (11)

- 11010004 – Middle White
- 11010008 – Current
- 11010012 – Strawberry
- 11020002 – Upper Arkansas
- 11020006 – Huerfano
- 11020010 – Purgatorie
- 11030001 – Middle Arkansas-Lake McKinney
- 11030005 – Pawnee
- 11030009 – Rattlesnake

- 11030013 – Middle Arkansas-Slate
- 11030017 – Upper Walnut River
- 11040003 – North Fork Cimarron
- 11040007 – Crooked
- 11050003 – Lower Cimarron
- 11060004 – Lower Salt Fork Arkansas
- 11070102 – Fall
- 11070106 – Caney
- 11070203 – Lower Cottonwood
- 11070207 – Spring
- 11080002 – Cimarron
- 11080006 – Upper Canadian-Ute Reservoir
- 11090102 – Punta de Agua
- 11090106 – Middle Canadian-Spring
- 11090204 – Lower Canadian
- 11100104 – Palo Duro (North Canadian Basin)
- 11100301 – Middle North Canadian
- 11110102 – Dirty-Greenleaf
- 11110201 – Frog-Mulberry
- 11110205 – Cadron
- 11120102 – Palo Duro (Red Headwaters Basin)
- 11120201 – Upper Salt Fork Red
- 11120303 – Lower North Fork Red
- 11130103 – North Pease
- 11130202 – Cache
- 11130206 – Wichita
- 11130210 – Lake Texoma
- 11130304 – Lower Washita
- 11140104 – Clear Boggy
- 11140108 – Mountain Fork
- 11140203 – Loggy Bayou
- 11140207 – Lower Red-Lake Latt
- 11140302 – Lower Sulpher
- 11140306 – Caddo Lake
- Other

Texas-Gulf Region (12)

- 12010003 – Lake Fork
- 12020002 – Middle Neches
- 12020006 – Village

- 12030103 – Elm Fork Trinity
- 12030107 – Cedar
- 12030202 – Lower Trinity-Kickapoo
- 12040103 – East Fork San Jacinto
- 12040203 – North Galveston Bay
- 12050002 – Blackwater Draw
- 12050006 – White
- 12060103 – Paint
- 12060202 – Middle Brazos-Lake Whitney
- 12070102 – Yegua
- 12070202 – Cowhouse
- 12080001 – Lost Draw
- 12080005 – Johnson Draw
- 12090101 – Middle Colorado-Elm
- 12090105 – Concho
- 12090109 – San Saba
- 12090203 – South Llano
- 12090301 – Lower Colorado-Cummins
- 12100101 – Lavaca
- 12100203 – San Marcos
- 12100303 – Lower San Antonio
- 12100403 – East San Antonio Bay
- 12100407 – Aransas
- 12110104 – Turkey
- 12110108 – Lower Frio
- 12110201 – North Corpus Christi Bay
- 12110205 – Baffin Bay
- Other

Rio Grande Region (13)

- 13010001 – Rio Grande Headwaters
- 13010005 – Conejos
- 13020202 – Jemez
- 13020206 – North Plains
- 13020210 – Jornada Del Muerto
- 13030103 – Jornada Draw
- 13040100 – Rio Grande-Fort Quitman
- 13040204 – Terlingua
- 13040208 – Reagan-Sanderson
- 13040212 – Amistad Reservoir

- 13050001 – Western Estancia
- 13060001 – Pecos Headwaters
- 13060005 – Arroyo Del Macho
- 13060009 – Rio Felix
- 13070003 – Toyah
- 13070007 – Landreth-Monument Draw
- 13070011 – Howard Draw
- 13080003 – International Falcon Reservoir
- Other

Upper Colorado Region (14)
- 14010002 – Blue
- 14020001 – East-Taylor
- 14020005 – Lower Gunnison
- 14030003 – San Miguel
- 14040102 – New Fork
- 14040106 – Upper Green-Flaming Gorge Reservoir
- 14040200 – Great Divide Closed Basin
- 14050004 – Muddy
- 14060001 – Lower Green-Diamond
- 14060005 – Lower Green-Desolation Canyon
- 14060009 – San Rafael
- 14070004 – Dirty Devil
- 14080101 – Upper San Juan
- 14080105 – Middle San Juan
- 14080202 – McElmo
- Other

Lower Colorado Region (15)
- 15010001 – Lower Colorado-Marble Canyon
- 15010005 – Lake Mead
- 15010009 – Fort Pearce Wash
- 15010013 – Meadow Valley Wash
- 15020002 – Upper Little Colorado
- 15020006 – Upper Peurco
- 15020010 – Chevelon Canyon
- 15020014 – Jadito Wash
- 15020018 – Moenkopi Wash
- 15030104 – Imperial Reservoir
- 15030108 – Yuma Desert

- 15030204 – Bill Williams
- 15040004 – San Francisco
- 15050100 – Middle Gila
- 15050301 – Upper Santa Cruz
- 15050305 – Aguirre Valley
- 15060103 – Upper Salt
- 15060201 – Big Chino-Williamson Valley
- 15070102 – Aqua Fria
- 15070202 – Tenmile Wash
- 15080103 – Tule Desert
- 15080303 – Cloverdale
- Other

Great Basin Region (16)
- 16010202 – Middle Bear
- 16020102 – Lower Weber
- 16020204 – Jordan
- 16020304 – Rush-Toole Valleys
- 16020308 – Northern Great Salt Lake Desert
- 16030002 – East Fork Sevier
- 16030006 – Excalante Desert
- 16040101 – Upper Humboldt
- 16040105 – Middle Humboldt
- 16040109 – Little Humboldt
- 16040204 – Massacre Lake
- 16050103 – Pyramid-Winnemucca Lakes
- 16050203 – Carson Desert
- 16050304 – Walker Lake
- 16060004 – Northern Big Smoky Valley
- 16060008 – Spring-Steptoe Valleys
- 16060011 – Ralston-Stone Cabin Valleys
- 16060015 – Ivanpah-Pahrump Valleys
- Other

Pacific Northwest Region (17)
- 17010104 – Flint-Rock
- 17010206 – North Fork Flathead
- 17010210 – Stillwater
- 17010214 – Pend Oreille Lake
- 17010302 – South Fork Coeur d'Alene

- 17010306 – Hangman
- 17020002 – Kettle
- 17020006 – Okanogan
- 17020010 – Upper Columbia-Entiat
- 17020014 – Banks Lake
- 17030002 – Naches
- 17040103 – Greys-Hobock
- 17040202 – Upper Henry's
- 17040206 – American Falls
- 170640210 – Raft
- 17040214 – Beaver-Camas
- 17040218 – Big Lost
- 17050101 – C.J. Strike Reservoir
- 17050105 – South Fork Owyhee
- 17050109 – Crooked-Rattlesnake
- 17050113 – South Fork Boise
- 17050117 – Lower Malheur
- 17050121 – Middle Fork Payette
- 17050201 – Brownlee Reservoir
- 17060102 – Imnaha
- 17060106 – Lower Grande Ronde
- 17060110 – Lower Snake
- 17060204 – Lemhi
- 17060208 – South Fork Salmon
- 17060302 – Lower Selway
- 17060306 – Clearwater
- 17070102 – Walla Walla
- 17070106 – Klickitat
- 17070204 – Lower John Day
- 17070304 – Upper Crooked
- 17080001 – Lower Columbia-Sandy
- 17080005 – Lower Cowlitz
- 17090003 – Upper Willamette
- 17090007 – Middle Willamette
- 17090011 – Clackamas
- 17100103 – Upper Chehalis
- 17100201 – Necanicum
- 17100205 – Alsea
- 17100302 – South Umpqua
- 17100306 – Sixes
- 17100310 – Lower Rogue

- 17110002 – Strait of Georgia
- 17110006 -Sauk
- 17110010 – Snoquaimie
- 17110014 – Puyallup
- 17110018 – Hood Canal
- 17120001 – Harney-Malheur Lakes
- 17120005 – Summer Lake
- 17120009 – Alvord Lake
- Other

California Region (18)

- 18010104 – Middle Fork Eel
- 18010108 – Big-Navarro-Garcia
- 18010202 – Sprague
- 18010206 – Upper Klamath
- 18010210 – Salmon
- 18020002 – Upper Pit
- 18020104 – Sacrament-Stone Corral
- 18020121 – North Fork Feather
- 18020126 – Upper Bear
- 18020152 – Cottonwood Creek
- 18020156 – Thomes Creek-Sacramento River
- 18020161 – Upper Coon-Upper Auburn
- 18030002 – South Fork Kern
- 18030006 – Upper Tule
- 18030012 – Tulare Lake Bed
- 18040006 – Upper San Joaquin
- 18040010 – Upper Stainislaus
- 18040014 – Panoche-San Luis Reservoir
- 18050003 – Coyote
- 18060002 – Pajaro
- 18060006 – Central Coastal
- 18060010 – Santa Ynez
- 18070101 – Ventural
- 18070105 – Los Angeles
- 18070202 – San Jacinto
- 18070302 – Santa Margarita
- 18080001 – Surprise Valley
- 18090102 – Crowley Lake
- 18090203 – Death Valley-Lower Amargosa
- 18090207 – Coyote-Cuddleback Lakes

- 18100202 – Carrizo Creek
- Other

Watershed Selection

Of the watersheds you selected above, for which ONE do you wish to provide answers in this survey? [Select one from previous choices]

Partnership Formation

Is your organization part of a watershed partnership or collaborative that works with other organizations, citizens, and governments to manage the watershed? – Yes – No

In what year did your organization join the partnership? [Text field]

How successful do you believe the partnership has been at achieving its goals? – Very Successful – Somewhat Successful – Neither Successful nor Unsuccessful – Somewhat Unsuccessful – Very Unsuccessful

Partnership Members

What social, private, nonprofit, or other nongovernmental organizations and associations are involved in the watershed? (Please list as many as you can.) [Text area] **What governmental organizations and agencies are involved in your watershed?** (Please list as many as you can.) [Text area]

Of the organizations you listed above, which are the top three that your organization most frequently interacts or communicates with? [Text area]

Decision-Making and Procedural Fairness

How significant is your organization's role in watershed management decisions? – Very Significant – Somewhat Significant – Neither Significant nor Insignificant – Somewhat Insignificant – Very Insignificant

Do you agree or disagree that the decision-making process is fair to all stakeholders? – Strongly Agree – Agree – Neither Agree nor Disagree – Disagree – Strongly Disagree

Do you agree or disagree that your organization's interests and concerns are adequately represented in the watershed management process? – Strongly Agree – Agree – Neither Agree nor Disagree – Disagree – Strongly Disagree

Even if your organization doesn't agree with some decisions, do you agree or disagree with the process used to make those decisions? – Strongly Agree – Agree – Neither Agree nor Disagree – Disagree – Strongly Disagree

If a collaborative partnership was available, would your organization join?
– Yes – No – I'm not sure

Constitutional Choice

Is there an official Watershed Plan that has been agreed to and adopted by watershed stakeholders?
- Yes
- No
- I don't know

Has your organization been responsible for producing any of the following? (Check all that apply.)
- Memorandum(s) of understanding for watershed partnership organizations
- Capital improvement programs
- Comprehensive land use plans
- A watershed plan
- A harbor management plan
- Mechanisms for conflict resolution
- Legal agreements
- Federal or state legislation
- Other

Which of the following activities does your organization develop in consultation with other organizations? (Check all that apply.)
- Priorities for habitat restoration
- Priorities for infrastructure investment
- Priorities for land acquisition
- Formal shared goals
- Formal shared policies
- Joint budgeting
- Agreements on formal monitoring protocols
- Agreements on formal water quality standards
- Agreements on formal water quantity standards
- Reports on progress toward environmental goals
- Reports on progress toward water quality or quantity enhancement activities
- Other

Collective Choice

In which of the following activities does your organization participate with other organizations in your watershed?

- Hiring staff to work in another organization
- Recruiting and training volunteers to support another organization
- Pooling financial resources for common activities
- Funding activities pursuant to another organization's priorities
- Staff that work in or support the work of another agency
- Other

In which of the following knowledge-sharing activities does your organization participate? (Check all that apply.)

- Joint research and fact finding
- Sharing informational resources and knowledge of current events
- Sharing knowledge of opportunities available to other resource users
- Inter-agency databases or database sharing (e.g., GIS)
- Joint development of technical information (e.g., monitoring data)
- Joint development of technical resources (e.g., equipment)
- Co-locating staff from different organizations
- Collecting information for another organization
- Ad hoc working groups with other organizations
- Informal working groups that meet regularly
- Formal collaborative organization that regularly meets
- Other

Operational Activities

Education and Outreach

In which of the following education and outreach activities does your organization participate? (Check all that apply.)

- Public education programs targeted at schools
- Public education programs for homeowners
- Public education programs for industry
- Public education programs for other resource users
- Public social events
- Special events and conferences for the public
- School curriculum development activities
- Training programs
- Special events and conferences for elected officials
- Special events and conferences for government administrators
- Other outreach programs and events
- Other

In how many of the above activities does your organization participate in an average year? [Dropdown menu]

Environmental Improvement Activities

In which of the following activities does your organization participate to improve environmental conditions? (Check all that apply.)

- Habitat restoration projects
- Clean-up activities
- Land acquisition
- Installing urban best management practices (BMPs)
- Installing agricultural BMPs
- Installing forestry BMPs
- Installing water flow BMPs
- Installing sewers to remove on-site sewage disposal systems (OSDSs)
- Upgrading OSDSs
- Other

In how many of the above activities does your organization participate in a given year? [Dropdown menu]

Please describe your organization's most recent activities related to the improvement of environmental conditions in your watershed. [Text area]

Water Management

Is an adaptive management approach used in your watershed to manage river or stream flow?

- Yes
- No
- I'm not sure

Do the rivers and streams in your watershed flow naturally (i.e., without impediments such as dams)?

- All of them
- Most of them
- Some of them
- None of them
- I'm not sure

If all of the water in your watershed does not flow naturally, is there a management approach that attempts to mimic natural flow (i.e., a natural flow regime)?

- Yes
- No
- I'm not sure

Collective Action Beliefs

Problem Severity

Concerning the overall health of your watershed, how severe do you think the problems are for each issue listed below? Please rate on a scale of 1 (Not Severe) to 10 (Very Severe):
- Water Quality: 1–10
- Water Quantity: 1–10
- Habitat Degradation: 1–10

Stakeholder Involvement

For your watershed to completely resolve its current problems, what percentage of stakeholders (citizens, businesses, and/or governments) do you estimate would have to change their behavior and activities?
0% (Small Percent) – 100% (Large Percent)

Scientific Understanding

On average, how adequate do you perceive the level of scientific understanding about the nature of the watershed and its resources? Rate on a scale of 1 (Inadequate) to 10 (Adequate):
- Scientific understanding of watershed and resources: 1–10

On average, how adequate do you perceive the level of scientific understanding about the causes of problems in your watershed? Rate on a scale of 1 (Inadequate) to 10 (Adequate):
- Scientific understanding of causes of problems: 1–10

Trust

Thinking about the range of contacts your organization has had with other stakeholders, how much do you trust these stakeholders to fulfill promises and obligations related to watershed health? Rate on a scale of 1 (Completely Distrust) to 10 (Completely Trust):
- Social/Nongovernmental Stakeholders: 1–10
- Governmental Stakeholders: 1–10

Conflict

Has there been significant conflict among watershed stakeholders in the past decade?
- Yes
- No
- I'm not sure

What is or has been the source of this conflict? [Text area]

Has the conflict been resolved?

- Yes
- No

How was the conflict resolved? Please list any organizations that were instrumental to the conflict's resolution. [Text area]

If the conflict has not been resolved, what do you think it would take to resolve it? [Text area]

Outcomes

Since your organization's involvement in the watershed, please indicate your beliefs about the following:

Do you believe water quality has improved?

- Yes
- No
- I'm not sure

Do you believe other ecological measures (e.g., water quantity, habitat restoration) have improved?

- Yes
- No
- I'm not sure

Do you believe water quality will continue to improve over time?

- Yes
- No
- I'm not sure

If water quality has improved, how or in what way(s)? [Text area]

If water quality has not improved, why do you believe this is the case? [Text area]

If other environmental measures have improved, how or in what ways? [Text area]

If other ecological or environmental measures have not improved, why do you believe this is the case? [Text area]

If you believe water quality will continue to improve, why? [Text area]

If you believe water quality will not improve over time, why? [Text area]

Partnership Sustainability

Do you believe the watershed partnership will last for the next five years or longer?

- Yes
- No
- I'm not sure

Why or why not? [Text area]

If the partnership is maintained in the long term, do you believe that its efforts will lead to a sustainable ecosystem?

- Yes
- No
- I'm not sure

Why or why not? [Text area]

Greatest Challenge

What do you believe is the greatest challenge facing your watershed today? [Text area]

Past Collaboration

Have there been previous attempts to form a partnership or collaborative to address watershed issues?

- Yes
- No
- I don't know

Do you consider previous attempt(s) at collaboration to be a success or failure?

- Success
- Failure

Why do you believe the past collaborative effort was a success? [Text area]

Why do you think the past collaborative effort was a failure? [Text area]

Comments

Do you have any additional comments about your organization's role in watershed management or about the watershed in general?

- Yes
- No

Please enter any additional comments below. [Text area]

Other Watersheds

Do you wish to answer questions about any other watersheds in which your organization participates?

- Yes
- No

Note: This survey was designed to collect data on watershed stakeholder participation, collaboration, and governance across the United States. It utilizes the Hydrologic Unit Code (HUC) system for watershed identification and is structured around theoretical frameworks of collective action and institutional analysis.

Index

Printed by Integrated Books International,
United States of America